GLOBAL
WARMING

Changing
Ecosystems

GLOBAL
WARMING

Changing Ecosystems

Effects of Global Warming

Julie Kerr Casper, Ph.D.

✓ Facts On File
An imprint of Infobase Publishing

CHANGING ECOSYSTEMS: Effects of Global Warming

Facts On File, Inc.
An imprint of Infobase Publishing
132 West 31st Street
New York NY 10001

Library of Congress Cataloging-in-Publication Data
Casper, Julie Kerr.
   Changing ecosystems : effects of global warming / Julie Kerr Casper.
      p. cm.—(Global warming)
   Includes bibliographical references and index.
   ISBN-13: 978-0-8160-7263-7
   ISBN-10: 0-8160-7263-9
   1. Ecology—Popular works. 2. Biotic communities—Popular works. 3. Global warming—Popular works. I. Title.
   QH541.13.C37 2009
   577.27'6—dc22   2009001411

Facts On File books are available at special discounts when purchased in bulk quantities for businesses, associations, institutions, or sales promotions. Please call our Special Sales Department in New York at (212) 967-8800 or (800) 322-8755.

You can find Facts On File on the World Wide Web at http://www.factsonfile.com

Text design by Erik Lindstrom
Illustrations by Sholto Ainslie, Richard Garratt, Accurate Art, and Lucidity Information Design
Photo research by the author
Composition by Hermitage Publishing Services
Cover printed by Bang Printing, Brainerd, MN
Book printed and bound by Bang Printing, Brainerd, MN
Date Printed: November 17, 2009
Printed in the United States of America

10 9 8 7 6 5 4 3 2 1

This book is printed on acid-free paper.

CONTENTS

PREFACE

We do not inherit the Earth from our ancestors—
we borrow it from our children.

This ancient Native American proverb and what it implies resonates today as it has become increasingly obvious that people's actions and interactions with the environment affect not only living conditions now, but also those of many generations to follow. Humans must address the effect they have on the Earth's climate and how their choices today will have an impact on future generations.

Many years ago, Mark Twain joked that "Everyone talks about the weather, but no one does anything about it." That is not true anymore. Humans are changing the world's climate and with it the local, regional, and global weather. Scientists tell us that "climate is what we expect, and weather is what we get." Climate change occurs when that average weather shifts over the long term in a specific location, a region, or the entire planet.

Global warming and climate change are urgent topics. They are discussed on the news, in conversations, and are even the subjects of horror movies. How much is fact? What does global warming mean to individuals? What should it mean?

The readers of this multivolume set—most of whom are today's middle and high school students—will be tomorrow's leaders and scientists. Global warming and its threats are real. As scientists unlock the mysteries of the past and analyze today's activities, they warn that future generations may be in jeopardy. There is now overwhelming evidence that human activities are changing the world's climate. For thousands of years, the Earth's atmosphere has changed very little; but today, there are problems in keeping the balance. Greenhouse gases are being added to the atmosphere at an alarming rate. Since the Industrial Revolution (late 18th, early 19th centuries), human activities from transportation, agriculture, fossil fuels, waste disposal and treatment, deforestation, power stations, land use, biomass burning, and industrial processes, among other things, have added to the concentrations of greenhouse gases.

These activities are changing the atmosphere more rapidly than humans have ever experienced before. Some people think that warming the Earth's atmosphere by a few degrees is harmless and could have no effect on them; but global warming is more than just a warming—or cooling—trend. Global warming could have far-reaching and unpredictable environmental, social, and economic consequences. The following demonstrates what a few degrees' change in the temperature can do.

The Earth experienced an ice age 13,000 years ago. Global temperatures then warmed up 8.3°F (5°C) and melted the vast ice sheets that covered much of the North American continent. Scientists today predict that average temperatures could rise 11.7°F (7°C) during this century alone. What will happen to the remaining glaciers and ice caps?

If the temperatures rise as leading scientists have predicted, less freshwater will be available—and already one-third of the world's population (about 2 billion people) suffer from a shortage of water. Lack of water will keep farmers from growing food. It will also permanently destroy sensitive fish and wildlife habitat. As the ocean levels rise, coastal lands and islands will be flooded and destroyed. Heat waves could kill tens of thousands of people. With warmer temperatures, outbreaks of diseases will spread and intensify. Plant pollen mold spores in the air will increase, affecting those with allergies. An increase in severe weather could result in hurricanes similar or even stronger than Katrina in 2005, which destroyed large areas of the southeastern United States.

Higher temperatures will cause other areas to dry out and become tinder for larger and more devastating wildfires that threaten forests, wildlife, and homes. If drought destroys the rain forests, the Earth's delicate oxygen and carbon balances will be harmed, affecting the water, air, vegetation, and all life.

Although the United States has been one of the largest contributors to global warming, it ranks far below countries and regions—such as Canada, Australia, and western Europe—in taking steps to fix the damage that has been done. Global Warming is a multivolume set that explores the concept that each person is a member of a global family who shares responsibility for fixing this problem. In fact, the only way to fix it is to work together toward a common goal. This seven-volume set covers all of the important climatic issues that need to be addressed in order to understand the problem, allowing the reader to build a solid foundation of knowledge and to use the information to help solve the critical issues in effective ways. The set includes the following volumes:

Climate Systems
Global Warming Trends
Global Warming Cycles
Changing Ecosystems
Greenhouse Gases
Fossil Fuels and Pollution
Climate Management

These volumes explore a multitude of topics—how climates change, learning from past ice ages, natural factors that trigger global warming on Earth, whether the Earth can expect another ice age in the future, how the Earth's climate is changing now, emergency preparedness in severe weather, projections for the future, and why climate affects everything people do from growing food, to heating homes, to using the Earth's natural resources, to new scientific discoveries. They look at the impact that rising sea levels will have on islands and other areas worldwide, how individual ecosystems will be affected, what humans will lose if rain forests are destroyed, how industrialization and pollution puts peoples' lives at risk, and the benefits of developing environmentally friendly energy resources.

The set also examines the exciting technology of computer modeling and how it has unlocked mysteries about past climate change and global warming and how it can predict the local, regional, and global climates of the future—the very things leaders of tomorrow need to know *today.*

We will know only what we are taught;
We will be taught only what others deem is important to know;
And we will learn to value that which is important.
—Native American proverb

ACKNOWLEDGMENTS

Global warming may be one of the most important issues you will have to make a decision on in your lifetime. The decisions you make on energy sources and daily conservation practices will determine not only the quality of your life but also that of your descendants.

I cannot stress enough how important it is to gain a good understanding of global warming: what it is, why it is happening, how it can be slowed down, why everybody is contributing to the problem, and why *everybody* needs to be an active part of the solution.

I would sincerely like to thank several of the federal agencies that research, educate, and actively take part in dealing with the global warming issue—in particular, the National Aeronautics and Space Administration (NASA), the National Oceanic and Atmospheric Administration (NOAA), the Environmental Protection Agency (EPA), and the U.S. Geological Survey (USGS)—for providing an abundance of resources and outreach programs on this important subject. I would especially like to acknowledge the years of leadership and research provided by Dr. James E. Hansen of NASA's Goddard Institute for Space Studies (GISS). His pioneering efforts over the past 20 years have enabled other scientists, researchers, and political leaders worldwide to better understand the scope of the scientific issues involved at a critical point in time when action must be taken before it is too late. I give special thanks to Al Gore and Arnold Schwarzenegger for their diligent efforts toward bringing the global warming issue so powerfully to the public's attention. I would also like to acknowledge and give thanks to the many wonderful universities across the United States, in England, Canada, and Australia, as well as to private organizations, such as the World Wildlife Fund and the Union of Concerned Scientists, that diligently strive to educate others and help toward finding a solution to this very real problem.

I want to give a huge thanks to my agent, Jodie Rhodes, for her assistance, guidance, and efforts, and also to Frank K. Darmstadt, my editor, for all his hard work, dedication, support, and helpful advice and attention to detail. His efforts in bringing this project to life were invaluable. Thanks also to the copyediting department for their assistance and the outstanding quality of their work, with a special thank you to Alexandra Lo Re for all of her input and enthusiasm toward this critical topic.

INTRODUCTION

The Earth is getting increasingly warmer—summers are growing hotter, glaciers are melting, sea levels are rising, and weather events are becoming more unpredictable. Human-induced global warming has only emerged as a serious issue in the past few decades, but climatologists have found evidence that humans are slowly changing the Earth's climate and environment. Although the change so far is not rapid, scientists expect it to begin accelerating. They predict that in the 21st century, the Earth will be hotter than it has been for most of the last 420,000 years.

Natural global warming occurs over time due to factors such as the relationships between the Earth's rotation, axis, position, and revolution around the Sun, as well as a result of major volcanic eruptions (what has traditionally been small increments of change over thousands of years). These types of gradual changes allow most species to survive by migration or adaptation. However, warming has increased dramatically during the last century at an unnatural rate, making specialists believe that the real cause of global warming today is human induced. Many activities humans are involved in—such as burning fossil fuels for energy and massive deforestation—are contributing to the atmospheric warming at an alarming rate. Experts believe that in the future enough human-induced damage will have been done to create severe problems in the distribution of species and their critical habitats, to increase the occurrence of severe weather events, to contribute to sea-level rise, and to trigger a host of health and quality-of-life issues that will affect everyone on Earth. Unfortunately, no ecosystem will escape the impact of human-induced global warming.

This volume in the Global Warming set looks at this serious issue and the far-reaching effects it is having right now, and will have in the future, on every ecosystem on Earth. It also explains why it is important for you—the reader—to understand the relevant issues now so that you can help solve this problem before it is too late and many species and habitats are gone forever.

In chapter 1, you learn about the effects of global warming on ecosystems—the current scientific findings, the observed effects, and the expected future effects. This chapter illustrates why even a few degrees of warming is a big deal.

Chapter 2 looks at the concept of biodiversity and just how vulnerable species can be, even in a well-established ecosystem. You see how disturbances and influences force a species to either adapt or die. This chapter also explains why changing land-use practices can wreak havoc on natural ecosystems, even though the change may be better or more convenient for humans.

Chapter 3 looks exclusively at the impacts of global warming on forest ecosystems. It classifies the forests of the world into three categories: temperate, boreal, and tropical; illustrates why they are the treasure troves of multitudes of natural resources; and examines just how global warming is threatening to destroy them. Some forest resources could be cures for diseases and may be destroyed before the cures are discovered. This chapter also shows you how humans are contributing to the problem and accelerating the destruction of forests, actually increasing their vulnerability to global warming.

Chapter 4 centers on warming's effects on the world's rangelands, grasslands, and prairies. Not often mentioned in the media when global warming is discussed, these ecosystems support an abundance of wildlife and serve as the source of much of the world's food. You will learn in this chapter just how critical these ecosystems are to every person on Earth and why it is important that the health of the ecosystems be maintained.

Chapter 5 takes you to the ends of the Earth—the polar regions. These regions are among the Earth's most fragile and are being affected the most drastically now. Wildlife, such as the magnificent polar bear, is facing extinction as the Arctic Ocean ice melts. Used as their hunting, breeding, and feeding grounds, without the ice, the polar bears simply cannot survive. Even though the polar regions seem far away, our use of fossil fuels is writing the bears' death sentences. This chapter also focuses on the dynamics of permafrost and why its melting will cause so much destruction. It also deals with shifting vegetation zones and what species may disappear because of them.

Chapter 6 focuses on desert ecosystems and what the specific threats are to them in a world of increased warming. It addresses the topic of drought, why and where it is increasing, and then touches on the process of desertification, where it can happen, who it can affect, and whether or not it is possible to manage. You will learn about deadly heat waves that have killed thousands of people already and what role the heat waves will play in the future. Finally, we will examine the frightening issue of wildfires and how fires, like those happening in Southern California, will only become more common in the years to come if global warming increases.

Chapter 7 addresses how warming affects the world's mountain ecosystems and the threats they face. It shows why the world's mountain regions are in danger and which parts have nowhere to go when the temperatures climb too high. It also gives you a perspective as to what global warming will do to the economy of these regions and why certain winter vacation resorts may not have many customers booking vacations in the future.

Chapter 8 provides a glimpse into the impacts currently felt by marine ecosystems—shoreline, deep ocean, temperate, tropical, and freshwater areas. The tragic fate of the world's spectacular reefs is explained, and we show what countries are attempting to do in order to stop their demise. It also focuses on the scarcity of the Earth's supply of accessible freshwater and what the consequences will be if we do not protect and manage it intelligently.

Finally, this volume looks at the future—what our options are, what we know based on what we have already seen happen to existing ecosystems now and in the past, and the many ways in which each one of us can help solve the problem that every ecosystem in the world is now experiencing. You will discover how to empower yourself and become eco-responsible. After all, we share this planet with a diverse array of life—and we need to keep it that way.

Signs and Effects of Global Warming

Global warming is one of the most controversial issues facing the world today. It is controversial not only because it involves very serious, far-reaching—global—issues that can permanently and negatively affect the world, but also because it involves the lifestyles and personal choices of everyone on Earth. While it is true that some people and ecosystems may suffer the negative effects more than others and some people may have to be willing to make larger lifestyle changes, everyone will have a responsibility and a stake in the outcome. If ecologically sound decisions are made, then life will be much better for not only today's population but also the generations of the future. If people choose not to make wise environmental choices today, then the future of upcoming generations will pay the price.

This chapter presents the current effects of global warming on ecosystems and what that means for the future, as well as the scientific findings of more than 2,500 scientists worldwide working together to better understand the pertinent aspects of global warming and what they mean to every individual on Earth. It then discusses global warming's present and potential effects on ecosystems if global warming continues unabated and why the Earth will become a very different place to try to adapt to in the future. In order to understand what certain parts of the world will be like in the face of global warming, it is first necessary to understand the concept of ecosystems and the related signs and effects of global warming.

THE EFFECTS OF GLOBAL WARMING ON ECOSYSTEMS

Although not every scientist worldwide may look at global warming in the same way, they do overwhelmingly agree that the Earth's atmosphere is getting warmer. Worldwide temperatures have risen more than 1°F (0.6°C) over the past century, and 17 of the past 20 years have been the hottest ever recorded. According to a special report issued by *Time* magazine on April 3, 2006, the *Intergovernmental Panel on Climate Change (IPCC),* in their third report, released in 2001, had analyzed data from the past two decades representing properties such as air and ocean temperatures and the habitat characteristics and patterns of wildlife. Examples of observed changes included "shrinkage of glaciers, thawing of permafrost, later freezing and earlier breakup of ice on rivers and lakes, lengthening of mid- to high-latitude growing seasons, poleward and altitudinal shifts of plant and animal ranges, declines of some plant and animal populations, and earlier flowering of trees, emergence of insects, and egg-laying in birds. Associations between changes in regional temperatures and observed changes in physical and biological systems have been documented in many aquatic, terrestrial, and marine environments."

The IPCC is an organized group of more than 2,500 climate experts from around the world that consolidates their most recent scientific findings every five to seven years into a single report, which is then presented to the world's political leaders. The IPCC was established in 1988 by the World Meteorological Organization (WMO) and the United Nations Environment Programme (UNEP) to specifically address the issue of global warming. As a result of their comprehensive analysis, they have determined that this steady warming has had a significant impact on at least 420 animal and plant species and also on natural processes. Furthermore, this has not just occurred in one geographical location but worldwide.

In the IPCC's fourth report, released in February 2007, they concluded that it is "very likely" (> 90 percent) that heat-trapping emissions from human activities have caused "most of the observed increase in globally averaged temperatures since the mid-20th century."

Also, in the February 2007 report, they concluded the following:

"Human induced warming over recent decades is already affecting many physical and biological processes on every continent. Nearly 90 percent of the 29,000 observational data series examined revealed changes consistent with the expected response to global warming, and the observed physical and biological responses have been the greatest in the regions that have warmed the most."

In these studies, scientists have been able to break down the natural and human-caused components in order to see how much of an effect humans have had. Human effects can include activities such as burning *fossil fuels,* agricultural practices, *deforestation,* industrial processes, the introduction of invasive plant or animal species, and various types of land-use change.

In many cases, scientists do not need to look very far to see the effects a warming world is having on the environment and the Earth's ecosystems. *Glaciers* worldwide are melting at an accelerated rate never seen before. The cap of ice on top of Kilimanjaro is rapidly disappearing, the glaciers of world-renowned Glacier National Park in the United States and Canada are melting and projected to be gone in the next few decades, and the glaciers in the European Alps are experiencing a similar fate.

In the world's *tropical* oceans, vast expanses of beautiful, brilliantly colored coral reefs are dying off as oceans slowly become too warm. Unable to survive the higher temperatures, the corals are undergoing a process called *bleaching* and are turning white and dying. In the Arctic, as temperatures climb, ice is melting at accelerated rates, leaving polar bears stranded, destroying their feeding and breeding grounds, and causing them to starve and drown. *Permafrost* is melting at accelerated

As the Earth warms under the influence of global warming, species that need cooler temperatures will attempt to migrate to cooler regions. They will move globally closer to the polar areas or locally up mountain ranges to higher elevations. *(Nature's Images)*

rates. As the ground thaws, it is disrupting the physical and chemical components of the ecosystem by causing the ground to shift and settle, toppling buildings and twisting roads and railroad tracks, as well as releasing *methane* gas into the atmosphere (another potent *greenhouse gas* responsible for global warming).

Weather patterns are also changing. *El Niño* events are triggering destructive weather in the eastern Pacific (in North and South America). There has been an increase in extreme weather events, such as hurricanes. Droughts have become more prevalent in some geographical areas, such as parts of Asia, Africa, Australia, and the American Southwest.

Animal and plant habitats have been disrupted, and, as temperatures continue to climb, there have been several documented migrations of individual species moving northward (toward the poles) or to higher elevations on individual mountain ranges. Migration patterns are also being affected, such as those already documented of beluga whales, butterflies, and polar bears. Spring is also arriving earlier in some areas, which is now influencing the timing of bird and fish migration, egg laying, leaf unfolding, and spring planting for agriculture. In fact, based on *satellite* imagery documentation of the Northern Hemisphere, growing seasons have steadily become longer since 1980.

While species have been faced with changing environments in the past and have been able to adapt in many cases, the IPCC climate change scientists view this current rate of change with alarm. They fully expect the magnitude of these changes to increase with the temperatures over the next century and beyond. The concern is that many species and ecosystems will not be able to adapt as rapidly as the effects of global warming will cause the environment to change. In addition, there will also be other disturbances, such as floods, insect infestations, and the spread of disease, wildfire, and drought. Any of these additional challenges can destroy a species or habitat. In particular, alpine (high mountain) and polar species are especially vulnerable to the effects of climate change because as species move northward (poleward) or higher on mountains, these species' habitats will shrink, leaving them with nowhere to go.

With so much evidence, most scientists no longer doubt that global warming is real, nor do they question the fact that humans are to blame. All it takes is a look at the air quality over significant population and industrial centers to begin to grasp the effect that humans can have on the environment.

Based on temperature records kept before the beginning of the *Industrial Revolution,* carbon dioxide (the most abundant greenhouse gas) in the *atmosphere* has increased 30 percent above those earlier levels. Not only are the levels higher, but they increase annually. According to the IPCC, at current conditions, by the year 2100, the average temperature is expected to increase between 2° and 11.5°F (1.1°–6.4°C)—an amount more than 50 percent higher than what was predicted only 50 years ago. Within the IPCC's predicted temperature range, at the lower end, storms would become more frequent and intense, droughts would be more severe, and coastal areas would be flooded by rising sea levels from melting glaciers and ice caps. There would be enough of a

One of the visible effects of the human-caused increase of CO_2 in the atmosphere can be seen as air pollution—also called smog—over the world's major cities. This photo is of Mexico City—the second-largest urban area in the world, with a current population of 18,131,000. *(Nancy A. Marley, Argonne National Laboratory)*

disruption that ecosystems worldwide would be thrown out of balance and altered. If, however, the temperature rise falls toward the higher end of the estimate, the results on ecosystems worldwide would be disastrous.

Sea levels could rise so much that entire islands of low elevation, such as the Maldives, could completely disappear. Other areas, such as the Nile Delta and much of the United States' coastal southeast (Florida, Louisiana, Mississippi), could become completely uninhabitable. *Climate* zones could shift, completely disrupting land-use practices. For instance, the current agricultural region of the Great Plains in the

United States (Nebraska, Oklahoma, Kansas) could be shifted to Canada. The southern portion of the United States could become more like Central or South America. Siberia would no longer be a frozen, desolate landscape. Parts of Africa could become dry, desolate wastelands. If this were to happen, it would have a severe impact on the production of agriculture. Areas currently equipped to produce agriculture would no longer be able to, and areas that were able, based on climate, may not have the financial *resources* or the proper soils. The ripple effect of these disruptions would be felt worldwide. Millions of people would be forced to migrate from newly uninhabitable regions to new areas where they could survive.

This would also affect public health. Rising seas would contaminate freshwater with salt water; there would be more heat-related illnesses and deaths; and disease-carrying rodents and insects, such as mice, rats, mosquitoes, and ticks, would spread diseases such as malaria, encephalitis, Lyme disease, and dengue fever.

Scientists of the IPCC agree that one of the most serious aspects of all this drastic change is that it is happening so fast. These changes are happening at a faster pace than the Earth has seen in the last 100 million years. While humans may be able to pick up and move to a new location, animals and their associated ecosystems cannot. The choices people make and the actions they take today will determine the fate of other life and their ecosystems tomorrow.

SCIENTIFIC FINDINGS

In order to understand climate trends and behavior, it is first necessary to study the climate history of the Earth's past, obtained from written documentation since record keeping of global temperatures began in 1867. Before that, consistent, reliable data are not directly available. Data can be obtained from sources such as ice cores, tree rings, and coral.

Once past evidence is put together, *climatologists* work at determining the cause and then try to project forward in order to determine what the outcome may be over the near term, mid term, and long term. Once specific contributions toward the climate's condition are identified, the information is loaded into powerful computer programs that try to model one of the most complex systems ever known—the Earth's climate system. Typically, when models are run, several different versions are run in order to provide a range of predictions, which is why estimates of future temperature rise are generally given within a range of values.

Invariably, the issue of climate change comes down to a key issue: Are humans involved, and, if so, by how much? Through the creation of climate models that take into account the interactions of the Earth's oceans and atmosphere, the only way scientists have been able to refine the models so that they accurately depict reality is by adding in the specific components of the climate caused by humans. This human footprint, as it is called, is the result of actions such as burning *fossil fuels,* destroying the world's forests, and changing land-use patterns. These activities are causing CO_2 to accumulate at accelerated rates in the atmosphere, where it then acts as an insulator and heats up the oceans, atmosphere, and the Earth's surface. These higher temperatures are serious enough that it is their presence in climate models that best duplicates the reality that scientists are measuring in the Earth's environment today.

In material that the IPCC has published in the two decades since its formation, the reports released in 2007 have been the strongest and most definitive yet as to the seriousness of global warming not only today but for future generations. A report issued by Dr. James Hansen of the National Aeronautics and Space Administration's Goddard Institute for Space Studies (NASA GISS) says that global warming in recent decades has taken global temperature to its highest level in the past millennium and that the Kyoto reductions will have little effect in the 21st century. Furthermore, the report states that "thirty Kyotos" may be needed to reduce warming to an acceptable level. Enough damage has been done at this point that there is no way to stop the effects of global warming now. The levels of CO_2 that have already been released into the atmosphere will continue to heat the Earth for decades to come. However, assertive action today can slow it down.

According to the IPCC, this past decade (1998–2007) is the warmest on record—the current global mean temperature (end of 2007) is estimated at 0.74°F (0.41°C) above the 1961–90 annual average of 57.2°F (14°C). In addition to the notable increase in temperature, 2007

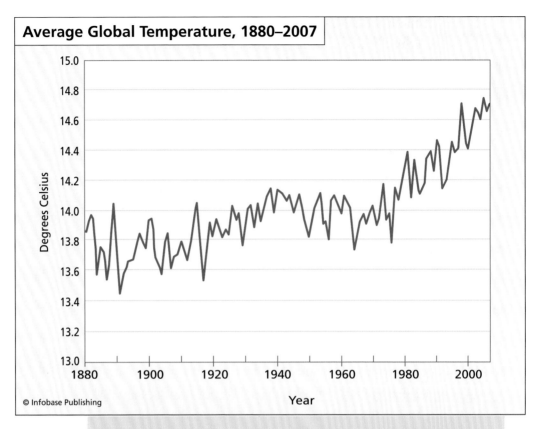

Average Global Temperature, 1880–2007

© Infobase Publishing

This chart depicts the average global temperatures recorded from 1880 to 2007. Although individual dips and spikes are present from year to year, the overall trend of temperature rise over time is apparent. *(NASA, GISS)*

also saw a record low Arctic sea ice level. In fact, enough ice melted in the Arctic that the Northwest Passage was ice free and passable by ships for the first time in recorded history.

The National Climatic Data Center (NCDC) states in its 2006 report that "during the past century, global surface temperatures have increased at a rate near 0.11°F (0.06°C) per decade, but this trend has increased to a rate roughly 0.32°F (0.18°C) per decade during the past 25 to 30 years." In addition, the NCDC's Preliminary Annual Report on the Climate of 2007, released on December 13, 2007, states the following points:

1. The global annual temperature for combined land and ocean surfaces for 2007 is expected to be near 58°F (14.4°C) and would be the fifth warmest since records began in 1880, and
2. the year 2007 is on pace to become one of the 10 warmest years for the *conterminous* United States since national records began in 1895.

These findings are based on a huge amount of data collected worldwide. On February 2, 2007, the IPCC's first volume in its Fourth Assessment Report states that "not only do the records show a warming trend during the past 50 years in temperatures taken on land, but also in ocean temperatures taken worldwide." This is significant, because it is these "warmer" ocean readings that assure scientists that the warming is not just occurring on land around cities, where a lot of heat is released from industry, traffic, and homes.

The NCDC has also stated that this period of worldwide warming has been pronounced in the conterminous United States. It states that the 2006 average annual temperature was the warmest on record, nearly identical to the record set in 1998 for the hottest year. The NCDC collects its data from a dense network of more than 1,200 points across the country at U.S. Historical Climatology Network (U.S. HCN) stations. When the data is collected, if it was from an urban location, the artificial heat effect caused by cities (called the urban heat island effect) has been removed so that the data is not biased if it is collected near large cities.

The IPCC, who until 2007 had been fairly conservative in its opinions and conclusions (partly because the data had to be accepted by more than 2,500 scientists on the panel), has concluded in its 2007 assessment that "warming of the climate system is unequivocal."

It concludes with "very high confidence (at least 90 percent chance of being correct) that the globally averaged net effect of human activities since 1750 (the start of the Industrial Revolution) has been one of warming" of the Earth's climate system.

OBSERVED AND EXPECTED EFFECTS ON ECOSYSTEMS

There are several pieces of physical evidence that scientists have already identified indicating that global warming is already in progress and affecting all the ecosystems on Earth. By monitoring the health of *ecological* conditions, scientists can see the effects climate change is having on the individual components that comprise the ecosystem. Because an ecosystem is such a tightly knit system of living things within their natural environment, if one component is affected, a ripple effect can be started, eventually endangering the entire ecosystem.

Polar and Ice-Related Changes

Of all the Earth's ecosystems, climate change in the polar regions is expected to be more rapid and more severe than anywhere else. If snow and ice are melted, this will greatly change the *albedo* of the environment. As darker surfaces increase, more sunlight will be absorbed, rapidly heating up the Earth's surface and atmosphere. In addition, as worldwide species continue to migrate northward under warming temperatures, thick dark vegetation will crowd areas that were once wide-open snowfields, also lowering the albedo.

According to the IPCC, the average annual temperature in the Arctic has increased by 1.67°F (1°C) over the past century—which equals a rate roughly twice as fast as the global average. Winter temperatures have been consistently 3.3°F (2°C) warmer over the past century. The effects of this warming have been seen in decreases in thickness and extent of sea ice, the melting of permafrost, and later freezing and earlier breakup dates of winter sea ice. Glaciers worldwide are also melting. There are glaciers on all the Earth's continents except Australia and at all latitudes from the Tropics to the polar regions. There is widespread evidence that glaciers are retreating in many areas of the world.

Because sea ice in the polar regions is breaking up earlier in the year, polar bears and walrus are already suffering. Their feeding and breeding grounds are disappearing, their territorial boundaries are gone, and they are dying. Early breakup is also affecting the hunting habits and lifestyles of northern native inhabitants, forcing many to abandon life-long rich, traditional cultures and relocate to other areas.

Fire and Drought

Over large areas of the Earth, nights have warmed up more than days have. In fact, since 1950, minimum temperatures on land have rapidly increased. An increase in warm temperatures will lead to increases in the number of heat waves that strike urban areas, which will cause more heat-related illnesses and deaths.

Global warming is causing a more intense hydrologic cycle with increased evaporation. The greater the evaporation rates, the more soils and vegetation will dry out. As temperatures rise and vegetation dries out, areas will become drier under droughtlike conditions, and wildfires will become more common. This occurred during the 2007 summer and fall in California. During this tragic event, more than 772 square miles (2,000 km^2) of land burned from Santa Barbara County to the U.S.-Mexico border. Nearly 100 people were injured, 9 died, and more than 1,500 homes were destroyed. Wildfires forced 265,000 residents to flee their homes. Enormous fires took place in 2008, resulting in a tremendous loss of property, habitat, and lives. To see details, visit http://www.firescope.org/fires.htm.

Biological Changes

Global warming also affects the occurrence and spread of disease. It makes large populations vulnerable if the pathogen is spread quickly. Warmer temperatures and more precipitation will help spread disease organisms from rodents and insects to larger areas. The world's undeveloped countries are expected to be hit the hardest.

Rising sea levels due to the melting of glaciers and ice caps are causing coral to lose the *symbiotic* algae that they must have for nutrition. This algae is also what gives them their beautiful, vibrant colors. When the algae die, the coral looks white and is referred to as bleached. It only takes a small increase in temperature (1.67°F or 1°C) above normal summer levels for periods of time as short as only two or three days to cause this reaction. In 1998, one of the hottest years on record, coral reefs worldwide experienced the most extensive bleaching ever recorded. Coral bleaching was reported in 60 countries and island nations in the Indian Ocean, Red Sea, Pacific Ocean, Persian Gulf, Mediterranean, and Caribbean.

Plant and animal species are located where the climatic factors (temperature, precipitation, soil condition, humidity, and airflow) enable them to thrive. If any of these characteristics change—as they do with global warming—then species will attempt to migrate. Whether they are successful or not depends on several factors: the rate of change, availability of acceptable habitat, a physical way to relocate to acceptable habitat, and avoidance of predators. If any of these factors works out wrong, the species can become threatened, endangered, or extinct.

According to NASA, based on data obtained through satellite observation over the past 20 years, areas in both North America and Eurasia have longer growing seasons associated with the buildup of greenhouse gases (GHGs) in the atmosphere. During the course of data collection, NASA scientists noticed dramatic changes in the timing of when leaves first appeared in the spring and when they fell off in the fall. By monitoring when things turn green, NASA scientists determined that in Eurasia the growing season is currently almost 18 days longer than it was two decades ago. Today, spring arrives a week earlier and autumn 10 days later than they did in the past.

Physical Changes

The rise in mean global surface temperature has caused spring to come earlier in many parts of the world. This has led to a longer growing season in middle and high-latitude areas. The effects of this are widespread—leaves come out earlier and stay longer, and breeding and migration patterns of wildlife are affected. For instance, in the northeastern United States, the frost-free season now begins approximately 11 days earlier than it did in the 1950s.

As worldwide temperatures climb, the *hydrologic cycle* will intensify, producing more intense phenomenon such as flooding, landslides, and erosion. The areas at highest risk are those at mid to high latitudes. Trends are already apparent in several polar locations in the Northern Hemisphere. According to P. Y. Groisman and D. R. Easterling of the NCDC, over the past few decades snowfall has increased about 20 percent over northern Canada and about 11 percent over Alaska.

An increase in snowfall has also been observed over China. According to T. R. Karl and R. W. Knight (also of the NCDC), observations for the last century indicate that extreme weather events (more than two inches [five cm] in 24 hours) in the United States have increased by

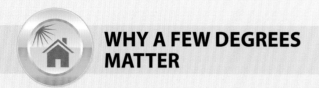

WHY A FEW DEGREES MATTER

When scientists predict that the Earth's atmosphere will get much warmer because of global warming, some people expect to hear that the temperature will get 10, 20, or 30 degrees warmer. Some have visions of sitting in a hot, steamy sauna, melting away. So, when climate scientists predict a temperature rise of 2–11.5°F (1.1–6.4°C), they are inclined to ask what the big deal is about global warming. It is just a few degrees, right?

Wrong.

Look at it this way: During the Earth's last ice age, the Earth was only about 6.7–10°F (4–6°C) cooler than it is today. Although it may not seem like much, these few degrees were responsible for blanketing huge areas of the Earth in thick layers of ice. It had such an impact on ecosystems that some animal species became extinct.

Thus, although a few degrees may seem trivial, the Earth's climate is so sensitive those few degrees can make a big difference. Scientists have already proven that temperatures are rising, making it something people cannot ignore now or in the future. The more people learn about the effects of global warming and how it can affect ecosystems worldwide, the more that can and must be done to slow down the changes. Society has progressed to a point where avoiding global warming is no longer an option, but learning how to slow it down, adapt, and have each person learn to do their part are important pieces of any solution.

about 20 percent. Increases in heavy rainfall have also been reported in Japan and northeastern Australia.

Tidal gauges are placed worldwide and mean sea level is monitored. Over the past century, global mean sea level has risen 4 to 10 inches (10 to 25 cm), with the average being 7 inches (18 cm). This rate is greater than what the average has been over the past few thousand years. The IPCC, although unsure how much ocean levels will rise in the next century, has projected that the rate will be at least two to four times the rate of the last century. The reason for the uncertainty is that the behavior of the Antarctic and Greenland ice sheets remains uncer-

Today is the time for action, and people can make a difference. It is key to understand the following:

- Global warming cannot be completely avoided any longer—it is already under way.
- The Northern Hemisphere is expected to warm up more than the Southern Hemisphere.
- Global warming is already having an impact on ecosystems, such as communities, forests, sea ice, permafrost, and wildlife habitats.
- Although some change in climate is normal, humans are causing the bulk of recent changes and rises in temperatures.
- It is humans who are adding GHGs at accelerated rates to the atmosphere, and there are measures that can be taken to help reduce climate change, such as using *renewable* energy and conserving precious natural resources and other environmentally friendly decisions.

Therefore, although a few degrees may not seem like a lot, it is enough to serve as a tipping point—a big enough influence on the climate that once reached will set into motion changes with global ramifications.

tain at this point. If their melting rate increases, future sea-level rise will most likely be on the larger side of the projection. Sea-level rise and coastal flooding are also governed by wind and pressure patterns, ocean circulation, and the characteristics of the coastline—whether there are coastal wetlands, beaches, islands, or other structures to act as barriers.

RESULTS OF GLOBAL WARMING ON ECOSYSTEMS

In a study on the effects of global warming on the Earth's ecosystems conducted by Chris Thomas, a conservation biologist at the University of Leeds in the United Kingdom, he states "Climate change now represents at least as great a threat to the number of species surviving on Earth as habitat destruction and modification." Thomas worked with a group of 18 scientists worldwide in the largest study of its type ever accomplished.

The end result of their study came down to this conclusion: "By 2050, rising temperatures, made more severe through human-induced input, such as the burning of fossil fuels, could send more than a million of Earth's land-dwelling plants and animals down the road to extinction."

The research team worked by themselves in six biodiversity-rich areas around the world, ranging from Australia to South Africa. As they gathered field data about species distribution and regional climate, they programmed the information into computer *climate models*. The purpose of the computer models was to simulate the direction and distance individual species would migrate in response to temperature and climate changes. Once all team members had collected their specific data, they combined it into a single model in order to understand the global concept.

Once the model had been carefully evaluated, it was determined that by the year 2050, at predicted global warming rates, 15 to 37 percent of the 1,103 species studied (165 to 408 species) could be at risk of extinction. When the study area was expanded to cover the entire Earth, the researchers estimated that worldwide more than a million species could begin to face extinction by 2050.

"This study makes clear that climate change is the biggest new extinction threat," said coauthor Lee Hannah of Conservation International (CI) in Washington, D.C.

"In some cases we found that there will no longer be anywhere climatically suitable for these species to live; in other cases they may be unable to reach distant regions where the climate will be suitable. Other species are expected to survive in much reduced areas, where they may then be at risk from other threats," said coauthor Guy Midgley of the National Botanical Institute in Cape Town, South Africa.

"Seeing the range of responses across all 1,103 species, it becomes obvious that we have a lot of work left to do before we can accurately predict what types of animals and plants are most at risk. This range of responses shows that species will not be able to move as whole biological communities, and that the typical natural communities we recognize today will probably not exist under future conditions. Figuring out what will replace them requires a lot of imagination," said coauthor Alison Cameron of the University of Leeds.

According to Chris Thomas, almost all future climate projections expect more warming and even more extinction between 2050 and 2100, and, even though projections are only made to 2100, temperatures will still keep going up and more warming will occur after that. This group of researchers say that taking action now to slow global warming is important to make sure that climate change ends up on the low end of the prediction in order to avoid "catastrophic extinctions."

Thomas also stated that because there may be a time lag between the climate changing and the last individual of a species dying off, the rapid reduction of greenhouse gas *emissions* may enable some species to survive. It is also important to keep in mind that although some species may be able to migrate successfully to a new location, some plant and animal species that live in high mountain or polar ecosystems cannot move farther to escape warming temperatures. Similarly, coral reef systems cannot just pick up and move to a new location. Long-established communities have remained where they are in order to survive.

Robert Puschendorf, a biologist at the University of Costa Rica, believes these estimates "might be optimistic." As global warming interacts with other factors such as habitat destruction, increase in invasive species, and the buildup of *carbon dioxide* in the environment, there may be more than a million species that face extinction.

In connection with Chris Thomas's study, Richard J. Ladle, et al., from Oxford University, published an article in *Nature* titled "Dangers of Crying Wolf Over Risk of Extinctions," bringing up an issue important to all scientific investigations. It cautioned that the scientific world needs to "learn how to deal with increasingly sensationalist mass media." They warn that policy-makers must be informed by a "balanced assessment of scientific knowledge and not popular perception created by commercially driven media. Departure from rational objectivity risks undermining public trust in the natural sciences and could play into the hands of antienvironmentalists. This places responsibilities on both scientists and journalists to ensure fair and accurate reporting of their work."

Ladle's report clarified that a large portion of the media incorrectly focused on the idea that more than a million species would definitely go extinct by 2050 when the study actually concluded that the extinctions will occur *eventually* and not in the next 50 years. They reported that 21 out of 29 articles quoting Thomas's study misinformed the public by inferring the species would be extinct by 2050.

Their chief concern is that the media, or anyone else, through unclear presentation, could inadvertently increase public cynicism and complacency about climate change and biodiversity loss. The resulting message is that when valuable studies, such as Chris Thomas's, are conducted and the results are released, in order to receive the best results from the scientific community, policy-makers, and the public in general, it is critical not to sensationalize or misrepresent the facts. Otherwise, it puts the real message in jeopardy and may even do more harm than good.

Both plant and animal species are at risk due to the effects of global warming. According to the World Wildlife Fund (WWF), the golden toad (*Bufo periglenes*) and the harlequin frog (*Atelopus varius*) of Costa Rica have already disappeared as a direct result of global warming. As different components of an ecosystem change, it can upset the natural balance in many ways. For instance, it can disrupt a species by having spring show up a week or two earlier. Over time, delicate balances have been set up between animals and the food they eat. If a certain animal relies on a specific food, but global warming has already caused the food to grow through its life cycle before the animal is ready to use it, then it will have a direct negative effect on that animal—the food will not be available when the animal needs it. That animal's health and existence may be threatened. Furthermore, this could cause a ripple effect in the *food chain,* having an impact on more than one species.

One example of this is when spring comes earlier than it has in the past. The timing of feeding for newly hatched birds may not correspond to the availability of worms or insects, impacting the fledglings' chances of survival. According to the WWF, climate records compared with long-term records of flowering and nesting times show a noticeable shift away from each other, depicting global warming trends. In Britain, flowering time and leaf-on records, which date back to 1736, have provided concrete evidence of climate-related seasonal shifts. Long-term trends toward earlier bird breeding, earlier spring migrant arrival, and later autumn departure dates have also been recorded in North America. Changes in migratory patterns have also been documented in Europe.

Once again, the WWF documents that climate change and global warming are currently affecting species in many ways. Animals and plants that require cooler temperatures in order to survive, which need to either migrate northward or higher in elevation in a mountain ecosystem, are already being documented in several places worldwide. This is occurring in the European Alps, in Queensland in Australia, and in the rain forests of Costa Rica. Fish in the North Sea have been documented migrating northward. Fish populations that used to inhabit areas around Cornwall, England, have migrated as far north as the Shetland and Orkney Islands. WWF global warming experts believe, based on this evidence, that "the impacts on species are becoming so significant that their movements can be used as an indicator of a warming world. They are silent witnesses of the rapid changes being inflicted on the Earth."

Species Threatened and Endangered by Global Warming	
polar bear	sea turtles
North Atlantic right whale	giant panda
marine turtles (multiple chelonian species)	
multiple bird species (mountain, island, wetland, arctic, antarctic, seabirds, migratory birds)	pikas
	wetland flora and fauna
snowy owls	salt wetland flora and fauna
mountain gorilla (Africa)	cloud forest amphibians
Andes spectacled bear	Bengal tiger

Habitats Threatened and Endangered by Global Warming	
coral reefs	mountain ecosystems
coastal wetlands	prairie wetlands
mangroves	
permafrost ecosystems	ice-edge ecosystems

In fact, these same scientists believe that global warming could begin causing extinctions of animal species in the near future because the heating caused by accelerated global warming has a severe impact on the Earth's many delicate ecosystems—both on the land and the species that live on it. Worldwide, there are species and habitats that have now been identified as being threatened and endangered due to the effects of global warming. These species and habitats can be seen in the table above and will be discussed in greater detail in later chapters of this book.

Because ecosystems can be altered to the point where the damage becomes irreversible and species must either adapt to survive or face extinction, it is critical that the issue of global warming be addressed and acted upon now before it is too late. The remainder of this book will illustrate why the best time to take positive action is today.

Ecosystems, Adaptation, and Extinction

This chapter introduces the concepts of *ecosystems*—what they are, why they exist where they do, how they function, what affects them, and the types of activities that can force them to change. It presents the concept of *biodiversity* and explains why it is so important. Next, it explores the issue of global warming and how it is currently affecting different plant and animal species around the world, as well as what will happen to them if global warming continues unchecked. It then examines key habitat preservation issues and how land-use change is having a huge impact on both ecosystems and the global warming dilemma. It is important for today and tomorrow's land managers to understand the cause-and-effect relationships of ecosystem health and global warming.

BIODIVERSITY AND ECOSYSTEMS

Diversity implies differences. A diverse habitat is composed of many different species. Having a wide diversity of life is beneficial to a habitat, because if one species were to die out, others will adapt and survive, perpetuating the habitat. Biodiversity describes the variety of life on Earth: It encompasses all the animals, plants, and microorganisms that exist on the planet, the genetic variety within these species, and the variety of ecosystems they inhabit. Ecosystem stands for ecological system. Biodiversity exists at the following three levels:

- Species diversity—this is the variety of species of animals, plants, and microorganisms on Earth or in a given area capable of reproducing fertile offspring.
- Genetic diversity—this is the variety of genetic characteristics found within a species and among different species. An example of genetic diversity in humans is the variation in eye color.
- Ecosystem diversity—this is the variety of natural systems found in a region, in a continent, or on the planet. A mountain, a prairie, and a wetland are three types of ecosystems.

Within each ecosystem, there are habitats that may also vary in size. A habitat is a place where a population lives in a community of living things that interact with the nonliving world around it to form an ecosystem. Most animals are only adapted to live in one or two habitats. For example, a bear can be seen in a temperate forest or in an alpine area, but not in a wetland or a desert. Saltwater fish—such as shark—could not live in a freshwater lake. Some animals migrate in the spring and the fall in search of warmer habitats with an abundance of food, such as humpback whales, which migrate to the Arctic in the summer and to Hawaii in the winter, and many species of birds.

The habitat must supply the needs of the organisms, such as food, water, appropriate temperature, oxygen, and minerals. If a population's needs are not met, it will move to a better habitat. Two different populations cannot occupy the same *niche* at the same time, however, because they would be competing against each other for survival. Because of this, the processes of competition, predation, cooperation, and *symbiosis* occur.

A habitat is the place where many species live naturally. Billions of different organisms live together in a habitat. Some, such as microbes, are so tiny that they can only be seen with a microscope. Others, like whales, horses, or giraffes, are huge. All species play a unique and significant role in making the habitat a healthy place to live.

Biodiversity is important in a habitat for several reasons. The different populations fill specific positions in the food chain. If a component in the food chain disappears, it upsets the ecological balance of the habitat and can ultimately destroy it.

The Earth's biodiversity has been in a constant state of change for as long as life has existed. Periodic natural events, such as meteor impacts, volcanoes, glaciers, climate change, drought, fire, and flooding, have disturbed ecosystems and led to the reduction or extinction of species, reducing biodiversity.

The reason global warming is having such a significant impact on ecosystems is that if a species cannot adapt or survive the changes in the environment brought on by global warming, it will become extinct. Once an animal leaves the food chain, the impact will be felt throughout the chain as other species attempt to adjust not only to warmer temperatures but also to the other physical and chemical effects of global warming. Warming conditions can affect both flora and fauna in a food web.

The more biodiversity there is—both within species and between species—the greater chance there is of that habitat surviving. Even if some species are destroyed, chances are good that others will survive when biodiversity is high. That is why biologists often measure the health of a habitat based on its biodiversity.

According to scientists at the Pew Center on Global Climate Change, the following are three major types of ecosystem changes that will occur as a result of the process of global warming:

1. Changes in the geographic distributions of vegetation types
2. Changes in ecosystem processes, such as productivity
3. Changes in the distributions and abundances of individual species

One of the major concerns about the effects of global warming during the next century is the adverse effect it will have on nature—the retreat of many of the Earth's forests because of climate stress, warming, and drought; the disappearance of delicate wetland areas due to increasing rates of sea-level rise; and the disappearance of many types of vegetation (trees and plants) that humans have become dependent on (such as sugar maples in New England), which may be gone by the end of the century under continued warming trends.

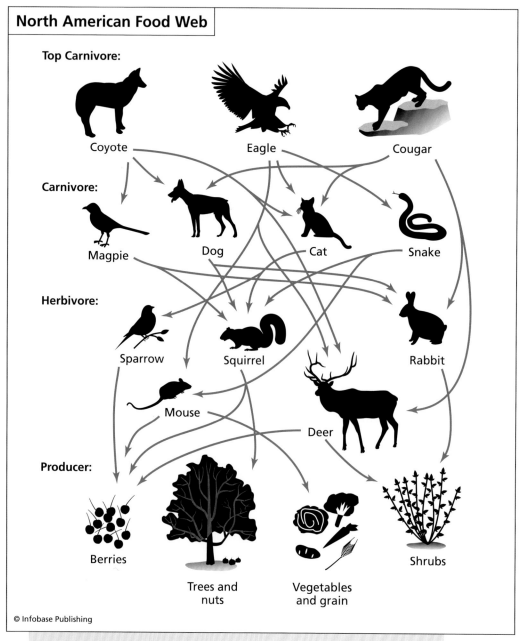

A typical North American food web

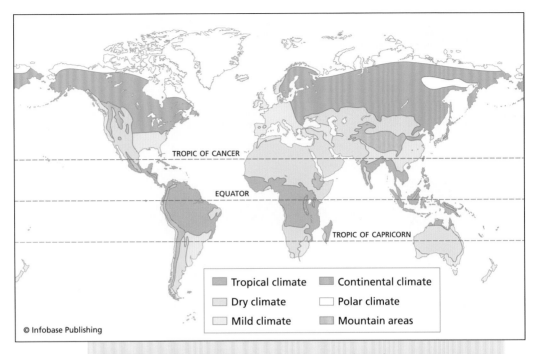

TROPIC OF CANCER

EQUATOR

TROPIC OF CAPRICORN

Tropical climate   Continental climate
Dry climate   Polar climate
Mild climate   Mountain areas

© Infobase Publishing

The Earth can be divided into climate zones based on temperature, moisture, and airflow.

Climate zones are expected to shift worldwide. When they do, it will have a devastating effect on biodiversity. When natural changes occur, they generally happen slowly over long periods of time, giving individual species a better opportunity to seek, find, and adjust to new habitats—a process in the past that has naturally occurred over a time period of thousands of years or more. With global warming, there is no luxury of time, and faster adaptations are demanded from species that they are unable to give.

In addition to the pressures created from a warming environment, additional pressures are being put on the landscape by humans through changes in land use, which is threatening the health and survival of many species. These human pressures, along with global warming, are causing significant losses to natural habitats and wildlife, such as to the Bengal tiger in the Sundarbans region of Bangladesh and India. Habitat loss is occurring in these areas because salinity levels are contaminating freshwater resources as the sea level continues to rise under the effects of global warming.

There are several types of ecosystems on Earth, each with its own unique characteristics. Descriptions of the environment, such as temperature and rainfall, are used to group habitats together. Habitats of similar climate and vegetation are called ecosystems. In different parts of the world, the same ecosystem may contain different species, but similar life-forms can always be identified. For example, a pine tree is a dominant form in a *temperate* forest, regardless of where the temperate forest is located. The Earth's ecosystems can be classified in several ways, including the following:

- Deserts
- *Tundras* and polar regions
- Boreal forests (also called taigas)
- Temperate grasslands
- Temperate forests
- Savannas
- Tropical rain forests
- Mountains
- Aquatic environments

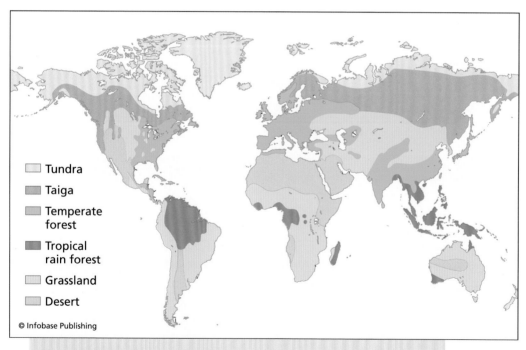

Tundra
Taiga
Temperate forest
Tropical rain forest
Grassland
Desert

© Infobase Publishing

The world's major ecosystems

These ecosystems consists of dynamic interactions between plants, animals, and microorganisms, which work together with their environment as a functional unit.

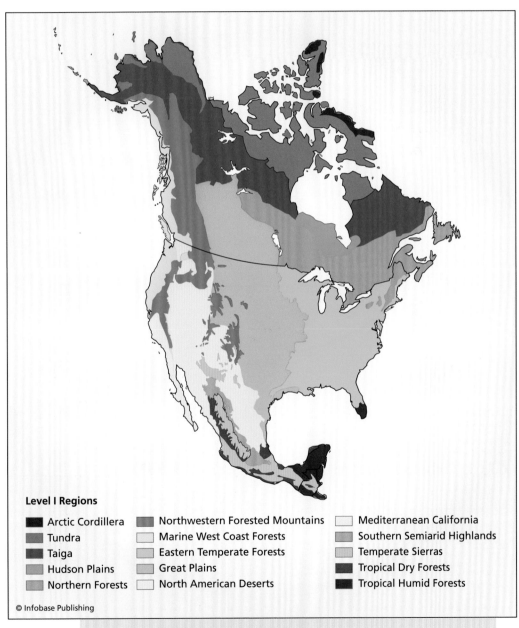

Level I Regions

Arctic Cordillera   Northwestern Forested Mountains   Mediterranean California
Tundra   Marine West Coast Forests   Southern Semiarid Highlands
Taiga   Eastern Temperate Forests   Temperate Sierras
Hudson Plains   Great Plains   Tropical Dry Forests
Northern Forests   North American Deserts   Tropical Humid Forests

© Infobase Publishing

The ecological regions of North America

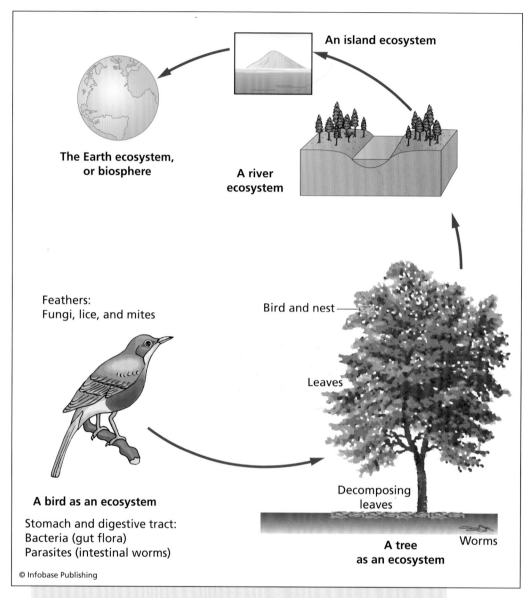

The concept of a nested ecosystem. Ecosystems can be several sizes, ranging from a broad area of land to a single bird.

Ecosystems exist at various levels, and some are nested within larger ecosystems, as depicted in the illustration. They change and move many times during the history of life on Earth. Global warming is one of the most crucial threats facing the world's ecosystems today.

The following summarizes the characteristics of the Earth's major biomes, but as global warming continues to raise the atmosphere's temperature and change the natural environment, the characteristics typical of these are being changed. Over time the changes will become irreversible.

Deserts

Deserts cover more than one-third of the Earth's land. They are some of the driest places on Earth and can be either hot or cold. Most lie between 20 to 30° North and South latitude. As a result of equatorial air falling toward the Earth's surface, this is where the great deserts of North Africa, southwestern North America, the Middle East, and Australia are located. Temperatures can climb as high as 120°F (50°C). Some cold deserts can drop to temperatures of -4°F (-20°C). The little rain that falls evaporates quickly. Even though the conditions are harsh, deserts provide a home to many animals.

Plant and animal species must be able to adapt to very demanding conditions, resulting in highly specialized vegetation (drought tolerant) and animals, both vertebrate and invertebrate. Desert soils generally contain abundant nutrients so that they only need water to become productive; they have little or no organic matter. Although not many large animals inhabit desert ecosystems (because they are not capable of storing sufficient water and enduring year-round), there is a notable variety of reptiles. Mammals that inhabit deserts are generally small.

Because all deserts are dry, they have significant daily temperature variations. Temperatures are high during the day because there is very little humidity in the atmosphere to block the Sun's electromagnetic energy from reaching the Earth's surface. Once the Sun goes down, however, the heat absorbed during the day quickly escapes back into space, making desert ecosystems very chilly at night.

Tundras and Polar Regions

These regions—located at the far northern and southern latitudes—are characterized by long, dark winters. They are the coldest biomes. Because they are located at the Earth's poles, the Sun never rises far above the horizon. Beyond the ice-covered Arctic, however, is the tundra, which supports a fairly large amount of life. There are about 1,700 kinds of plants in the Arctic and subarctic. Each summer, the vegetation that appears during the short growing season attracts many birds and other animals.

Tundra covers about one-fifth of the Earth's land surface. Because it is so cold (it can reach -50°F (-46°C in the winter), trees do not grow there; instead, it is an ecosystem composed of low-growing plant life and wildflowers, giving it the appearance of a frozen prairie.

During the tundra's brief summer, insects hatch out of eggs that were frozen in the topsoil, which provide food for thousands of migrating birds, as well as hawks, ravens, and ptarmigans. It also provides habitat for caribou, mink, weasel, arctic hare, wolf, wolverine, reindeer, and brown bear. Polar bear, walrus, and arctic fox inhabit the pack ice and coastal areas of the polar regions.

Boreal Forests

The boreal forest—also called the taiga—is an extremely large biome covering 6,800 miles (4,506 km) across the Northern Hemisphere. It also exists high on mountain ranges such as the Rocky Mountains and Appalachians in the United States and the Alps in Europe.

The boreal forest is characterized by a cold climate, scarce rainfall, and a short growing season. When boreal forests are located in the interior portions of continents, the temperatures can get colder than the polar regions farther north. Interior continental locations are colder because they do not have the benefit of nearby warmer ocean air to modify the climate. Frost covers the areas most of the year, which causes much of the ground to be permanently covered in water because the frost keeps the water from being able to drain away. The severity of winter limits the diversity of animal life.

Many of the animals do not live in the boreal forest the entire year. Birds only migrate into the area during the summer, moving south to a warmer climate for the rest of the year. Animals that do live in the boreal forest year-round have adapted to life in this harsh climate. Insects, such as mosquitoes, gnats, and midges, are abundant during spring and summer, serving as food for migratory birds.

Temperate Grasslands

Grasslands of Asia are called steppes, those of South America are called pampas and campos, and those of North America are called prairies. Many animals live in these areas. Different animals are found in different parts of the grasslands based on the annual rainfall of the area.

The North American prairie is divided into three categories: (1) short grass, where grass stays below 20 inches (51 cm) tall; (2) mixed grass, where grass ranges from 20–60 inches (51–152 cm) tall; and (3) eastern tall grass, where grass can grow as high as 10 feet (3 m).

The Asian steppe is divided into an eastern and western range. The eastern steppe is higher and drier. It can also be divided into western forest steppe (this area contains pine, oak, and birch trees), open steppe (where trees are scarce), and southern semidesert (where desert vegetation grows).

The various landforms—such as marshes, streams, rocky slopes, and rock outcrops—provide niches for various kinds of animals. Muskrat and water snakes are found in water bodies such as creeks and rivers; frogs, crickets, and water bugs are found in the shallow tide pools. The biome is also home to animals that graze (such as cattle and llamas), burrowing animals, and many different kinds of insects.

Wildfires play an important role in grasslands. They keep the areas free of trees and shrubs. As an example of *adaptation,* plants in the grasslands have adapted to the wildfires and need them in order to keep healthy and produce new vegetation in the spring.

Temperate Forests

Temperate forests used to cover large areas of eastern North America, Europe, and Asia. This biome is found in the middle latitudes and is seasonal, having warm summers and cold winters. The trees are largely deciduous and change to beautiful colors in the fall. This is one of the most altered biomes on Earth, because this is where most of the world's population lives. Temperate forests cover eastern North America, northeastern Asia, and western and central Europe. Human activity, however, has eliminated much of the forest areas. Today, only isolated regions of the forests can be found in the Northern Hemisphere.

The tree habitats support many animals. The forests of North America have five times as many species of trees as the forests in Europe. These forests provide homes for rare animals, such as the ivory-billed and red-cockaded woodpeckers. Animals that live in the forest clearings include birds and insects, such as the speckled wood butterfly.

Savannas

Savanna is of Spanish origin and means "treeless plain." Savannas are located between the tropic of Cancer and the tropic of Capricorn—the region of Earth extending from 23.5° North to 23.5° South latitude of the equator. They are found in Africa, India, and the northern part of South America. Climate is the most important factor in creating a savanna. They are large expanses of grass with only scattered trees. Another characteristic of a savanna is low rainfall (although not as low as a desert). Massive herds of grazing animals, such as zebra, giraffe, impala, gazelle, and wildebeest—along with predators such as lion, leopard, and cheetah—inhabit the savanna. Wild fruit trees provide ample food for birds and animals. Temperatures are high because savannas are located in the tropical latitudes. The rain falls mostly during two times of year. Savannas support a wide amount of biodiversity.

Tropical Rain Forests

The *tropical* rain forest is the richest and most diverse biome in the world because of its warm, damp climate. A total of more than 1.5 million species of plants and animals live in the rain forest.

Animals that live in the highest trees almost never visit the ground. These include insect-eating birds such as the hornbill and toucan. The main canopy of the forest traps moisture and shields the rain forest from the wind. This is the zone where the majority of animal species are found—scientists have discovered 600 different species of beetle in one canopy.

The tropical rain forests face a serious threat from human interaction and land-use practices that contribute to global warming. Deforestation has become a huge problem and is one of the leading causes of global warming. The rain forests store huge amounts of carbon dioxide (CO_2). As forests are cut down and burned—the equivalent of 50 football fields are destroyed every minute—the CO_2 that has been stored in the trees is released to the atmosphere to add to the rising CO_2 levels, causing global warming. Furthermore, the *carbon sink* resource (the tree) is no longer available to store additional CO_2. Both of these issues are critical to global warming.

Mountains

Occupying one-fifth of the world's land surface, one thing that distinguishes mountains from other ecological systems is that each mountain can have its own distinct climate and collection of plants and animals. Some mountains are permanently capped with snow. Mountains are usually wetter than lower areas. The side of the mountain that faces the Sun also has an impact on the vegetation that grows there, which also helps determine which animals will live there. One side of the mountain can be warm and damp, the other cold and windy. All mountain ecosystems have one thing in common—rapid changes in altitude, climate, soil, and vegetation over very short distances. Mountain ranges support a high degree of biodiversity. A few animals can survive in the harsh conditions of the mountains. Larger animals are usually agile, such as the mountain goat, and many animals adapt survival mechanisms to live in this environment.

Aquatic Environments

Water makes up the largest biosphere—oceans cover 75 percent of the Earth's surface. Characteristics such as light, temperature, pressure, and nutrients all affect their variety and abundance. In general, the number of species is greatest in the equatorial regions and least near the poles.

The transparency of water allows light to travel through it. Phytoplankton and many other organisms use light to make food through the process of photosynthesis. The surface waters, rich in nutrients, provide 12.6 million tons a year to fisheries. Coral reefs are highly diverse ecosystems that exist in warm, shallow waters. They provide habitat for corals, microorganisms, invertebrates, fishes, sea urchins, octopuses, eels, and starfish.

Estuaries are areas where freshwater streams or rivers merge with the ocean and provide a home for worms, oysters, crabs, and water-

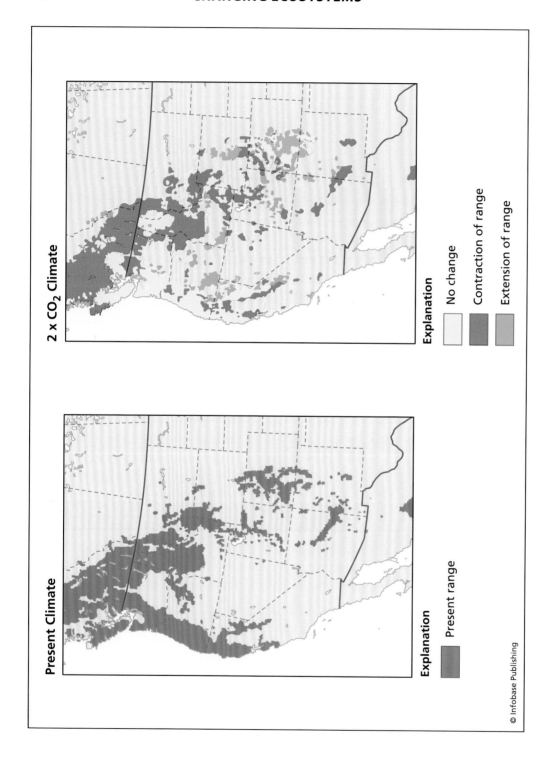

2 x CO₂ Climate

Present Climate

Explanation

No change

Contraction of range

Extension of range

Explanation

Present range

fowl. Freshwater regions—such as ponds, lakes, streams, rivers, and wetlands—have a low salt *concentration* (usually less than 1 percent). The species in these regions have adjusted to the low salt content and would not be able to survive in areas of higher salt concentration, such as the ocean.

Wetlands have the highest species diversity of all ecosystems. Many species of amphibians, reptiles, birds (such as ducks), and furbearers can be found in wetlands. Some wetlands have a higher salt content and provide habitat for shrimp and shellfish.

THE EFFECTS OF CLIMATE CHANGE ON ECOSYSTEMS

One way that scientists gain understanding of how global warming will affect ecosystems is to analyze the effects of past climate on paleoecosystems. A paleoecosystem is an ecosystem that existed in a former geologic time period. By relating vegetative cover to past climates, models can be developed. Once a reliable model is created, input variables such as CO_2 can be varied and the results analyzed. An example of what the effect would be like on populations of Douglas fir (*Pseudotsuga menziesii*) in the northwestern United States if the CO_2 content in the atmosphere were double what it was before the Industrial Revolution was modeled by the U.S. Geological Survey (USGS) and is shown in the illustration that follows.

In a study conducted by the U.S. Global Change Research Program (USGCRP) in Washington, D.C., first in 2001, then updated in 2004, in trying to predict what the effects of future climate change would have on ecosystems, they concluded that "climate change has the potential to

(opposite page) Distribution of Douglas fir in the northwestern United States. The illustration at left depicts present Douglas fir habitat. The illustration at right depicts the distribution with double the CO_2 in the atmosphere. Green areas show where the Douglas fir is today and could continue to live in the future. Red indicates the areas where the Douglas fir exists now but would no longer exist; blue indicates where the Douglas fir does not exist today but could be found in the future. *(USGS)*

affect the structure, function, and regional distribution of ecosystems, and thereby affect the goods and services they provide."

They based their conclusions on a modeling and analysis project they conducted called the Vegetation/Ecosystem Modeling and Analysis Project (VEMAP). This project was used to generate future ecosystem scenarios for the conterminous United States (excluding Alaska and Hawaii) based on model-simulated responses to both the Canadian and Hadley scenarios (see following paragraphs) of climate change.

The VEMAP was subsequently used in a validation exercise for a Dynamic Global Vegetation Model (DGVM) by Oregon State University and the U.S. Forest Service in 2008. Their MC1-DGVM was used as the input data in both the VEMAP and VINCERA (Vulnerability and Impacts of North American Forests to Climate Change: Ecosystem Responses and Adaptation). Their MC1 was run on both the VEMAP and VINCERA climate and soil input data to document how a change in the inputs can affect model outcome. The *simulation* results under the two sets of future climate scenarios were compared to see how different inputs can affect vegetation distribution and carbon budget projections. The results indicated that "under all future scenarios, the interior west of the United States becomes woodier as warmer temperatures and available moisture allow trees to get established in grasslands areas. Concurrently, warmer and drier weather causes the eastern deciduous and mixed forests to shift to a more open canopy woodland or savanna-type while boreal forests disappear almost entirely from the Great Lakes area by the end of the 21st century. While under VEMAP scenarios the model simulated large increases in carbon storage in a future woodier west, the drier VINCERA scenarios accounted for large carbon losses in the east and only moderate gains in the west. But under all future climate scenarios, the total area burned by wildfires increased." The similarities of the two models served to validate the VEMAP project.

The Hadley model was developed by the Hadley Centre for Climate Prediction and Research in England. Also referred to as the Met Office Hadley Centre for Climate Change, it is based at the headquarters of the Met Office in Exeter. It is the key institution in the United Kingdom for climate research. It is currently involved not only with understanding the physical, chemical, and biological processes within the climate system, but also with developing working models to explain current phenomena and to predict future climate change. It also monitors global and national climate variability and change and strives to determine the causes of the fluctuations. The Canadian climate model was developed by the Canadian Centre for Climate Modelling and Analysis (CCCma). The CCCma is a division of the Climate Research Branch of Environment Canada based out of the University of Victoria, Victoria, British Columbia. Its specific focus is on climate change and modeling. In the past nine years, the CCCma has produced three atmospheric and three atmospheric/oceanic general circulation models (GCMs), making them one of the international leaders in climate change research. What they found was that over the next few decades climate change in the United States will most likely lead to increased plant productivity as a result of increasing levels of CO_2 in the atmosphere. There will also be an increase in terrestrial carbon storage (such as in trees) for many parts of the country, especially the areas that become warmer and wetter (more humid and tropical-like). The southeast will most likely see reduced productivity and, therefore, a decrease in *carbon* storage (because of less vegetation growth).

By the end of the 21st century, many areas of the country will have experienced changes in the distribution of vegetation. Wetter areas will see the growth of more trees; drier areas (such as the southeast) will have drier soils that will cause forested areas to die off and be replaced by savanna/grassland ecosystems.

Modeling the vegetation evolution and adaptation is more difficult. The study focused on two time periods: 2025–2034 (near term) and 2090–2099 (long term). In the near term, biogeochemical changes are expected to dominate the ecological responses. Biogeochemical responses include changes based on the natural cycles of carbon, nutrients (such as *nitrogen*), and water. The responses are affected by changing environmental conditions such as temperature, precipitation, solar *radiation*, soil texture, and atmospheric CO_2. It is these natural cycles that affect carbon capture by plants with *photosynthesis,* soil nitrogen processes, and water transfer (*evapotranspiration* is *evaporation* and transpiration). These biogeochemical factors are what influence the production of vegetation.

In the results from the near-term biogeochemical model, the scientists concluded that there would be an increase in CO_2. They estimate that currently the average carbon storage rate is 66/Tg/yr (Tg = teragram, or 10^{12} grams). The Hadley model predicted that carbon storage rates by 2025–2034 would increase to 117/Tg/yr (almost a 100 percent increase over today's levels). The Canadian model estimated CO_2 to increase 96 Tg/yr. The Canadian model projected that the southeastern ecosystems will lose carbon in the near term because they predict the climate there will become hot and dry.

The biogeography models look at the changing landscape based on changes in CO_2, evapotranspiration, vegetation establishment, and competition between species, growth rates, and life cycle/mortality rates. In this model, scientists at both the Hadley and Canadian Centre for Climate Change agree that vegetation will be able to freely move from one location to another (under the constraints of agricultural and urban development). Changes in vegetation distribution will vary from region to region as follows:

- Northeast: Forests remain the dominant natural vegetation, but forest mixes will change (e.g., winter deciduous forests will replace mixed conifer-broadleaf forests). There will also be some increase in savannas and wetlands.
- Southeast: Forests remain the dominant ecosystem, but mixes change. Savannas and grasslands encroach on forests, especially toward the end of the 21st century. Drought and wildfires contribute significantly to forest destruction.
- Midwest: Forests remain the dominant land cover, but changes in species type occur. There will be a modest expansion of savannas and grasslands.
- Great Plains: Slight increase in woody vegetation.
- West: The areas of desert ecosystems shrink, and forest ecosystems grow.
- Northwest: Forested areas grow slightly.

A separate study conducted by the Canadian government predicts the following ecosystem changes as a result of changing climate over the next 100 years:

- Coastlines:
    - Flooding and erosion in coastal regions
    - Sea-level rise (threatening urban developments)
- Forests:
    - Increase in pests (insect invasions)
    - Increased levels of drought and wildfire
- Plants and animals:
    - Warmer temperatures could make water supplies more scarce, having a negative impact on plants and animals, not giving them time to adjust.
- Crops:
    - In some areas, warmer climate may allow a three- to five-week extension of the frost-free season, which could benefit commercial agriculture.
    - In other areas, drier soils and lack of water will have a negative impact on agricultural productivity.
- Wells:
    - The quality and quantity of drinking water may be threatened by increasing drought.
- Harsh weather:
    - Winter storms, floods, drought, heat waves, and tornadoes could become more frequent and severe.
- Fisheries:
    - Populations and ranges of species sensitive to changes in water temperature will be negatively affected.
    - Salmon harvests will be lower in the Pacific.
    - Changes in ocean currents may have a negative impact on the fisheries in the Atlantic.
- Lakes and rivers:
    - Water levels will decline under the influence of drought, negatively affecting drinking water quality.
    - Use of lakes for transportation, recreation, and fishing, and the ability to generate electricity (hydroelectric power) may be curtailed under droughtlike conditions.
    - Other areas that may have an increase in precipitation may experience flooding, rising sea levels, and severe storms.

In February 2005, the Met Office in Exeter, England, issued a report titled "Avoiding Dangerous Climate Change." The objective of the study was to determine what levels of CO_2 were considered the tipping point for dangerous climate change with harmful effects on ecosystems and what actions could be taken now to avoid such an outcome. In the report, then prime minister Tony Blair stated: "It is now plain that the emission of greenhouse gases . . . is causing global warming at a rate that is unsustainable."

Environment Secretary Margaret Beckett stated, "The report's conclusions would be a shock to many people. The thing that is perhaps not so familiar to members of the public . . . is this notion that we could come to a tipping point where change could be irreversible. We're not talking about it happening over five minutes, of course, maybe over a thousand years, but it's the irreversibility that I think brings it home to people."

The report, published by the British government, says there is only a small chance of greenhouse gas emissions being kept below "dangerous" levels. It warns that the Greenland ice sheet could melt, causing sea levels to rise by 23 feet (7 m) over the next 1,000 years. It also warns that developing countries will be the hardest hit.

The report also states, based on the vulnerability of many of the world's ecosystems, that the European Union (EU) has adopted a target of preventing an increase in global temperature of more than 3.3°F (2°C). Some believe even that may be too high. The report states: "Above two degrees, the risks increase very substantially," with "potentially large numbers of extinctions" and "major increases in hunger and water shortage risks . . . particularly in developing countries."

In order to meet their goals, British scientists have advised that CO_2 levels should be stabilized at *450 parts per million (ppm)* or below. Currently the atmosphere contains 380 ppm. In response to this, the British government's chief scientific adviser, Sir David King, said that was unlikely to happen. He stated, "We're going to be at 400 ppm in 10 years' time. I predict that without any delight in saying it."

Myles Allen, an expert on atmospheric physics at Oxford University, said that: "Assessing a 'safe level' of carbon dioxide in the atmosphere was 'a bit like asking a doctor what's a safe number of cigarettes to smoke per day.'"

The report does conclude, however, that there are technological options available to reduce CO_2 emissions that will need to be used. The study also concluded that the biggest obstacles involved with using these new technologies, along with renewable resources of energy and "clean coal," are the current economic investments and traditionally strong bond to the oil industry, cultural attitudes that oppose change, and simple lack of awareness by many people. Various conservation organizations currently involved in the battle against global warming, such as the Union of Concerned Scientists (UCS), the Defenders of Wildlife, and the World Wildlife Fund (WWF), also support these ideas.

IMPACTS, VULNERABILITY, AND ADAPTATION

One of the most serious constraints to animal and plant adaptation is interrupted migration. Migration for many species, such as geese, elk, salmon, leatherback turtles, wildebeest, and monarch butterflies, is a natural annual act. As global warming affects ecosystems worldwide, many species unable to survive in the new climatic conditions of the geographic areas where they have always lived will need to migrate to areas with new climates in which they can survive. There are several potential problems with this, however. Under the effects of global warming, the environment may change faster than a species can adapt. In other cases, existing land use—such as heavily urbanized areas—may have a negative impact on wildlife habitat. Wide-ranging wildlife species need secure core habitat where human activity is limited, ecosystem functions are still intact, and wildlife populations are able to flourish. If they do not have this, their long-term health and survival will deteriorate. In this situation, it is important that corridors connecting core areas are established before the effects of global warming are felt. As shown in the illustration, the "connectivity or roaming space between core habitat areas is maintained through a corridor," allowing the wildlife to freely move between core areas, as necessary. A buffer zone surrounds the critical core areas to maintain the integrity of the ecosystem. The purpose of the core areas, corridors, and buffers is to limit human activities so that wildlife can exist under optimal biodiversity.

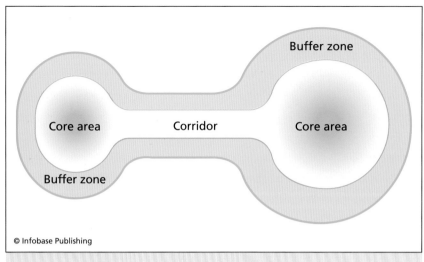

© Infobase Publishing

Wildlife corridors connect key core habitat areas.

Most ecosystems on Earth will be affected. Some of the most vulnerable include arctic fauna, such as polar bears and emperor penguins, and many salt wetland flora and fauna species, and species that inhabit lowland areas susceptible to flooding. In addition, species that rely on cold climate conditions will also be hit hard.

As presented in a report issued by LiveScience in June 2005, many changes in plants and animals will be seen as a result of this expected unnatural climate change. Specifically, they note that reindeer herds, typical in high latitudes in the Northern Hemisphere, will most likely disappear from large portions of their current range because of human-caused global warming. Several animal cycles have already become disrupted. For instance:

- Canadian red squirrels are now breeding 18 days earlier.
- Marmots are coming out of winter hibernation three weeks earlier than they did 30 years ago.
- Red fox are migrating northward, encroaching on the ecosystems traditionally occupied by arctic fox, causing a fight over natural resources for survival.
- Polar bears are thinner and less healthy than they were 20 years ago.

- Elephant seal pups are starving because their prey is migrating to cooler waters.
- Loggerhead sea turtles lay their eggs 10 days earlier than they did 15 years ago.
- Tidal organisms (rock barnacles, mollusks) are migrating northward.
- Many fish species are migrating northward in search of cooler water.
- The diet of some songbirds is changing. Some are now avoiding insects that eat leaves exposed to high levels of CO_2.
- North American tree swallows are laying their eggs about nine days earlier than they did 40 years ago.
- Some plants are currently growing in areas where they could not before because rising temperatures have provided greater amounts of heat, sunlight, and water.
- American flowering plants (e.g., columbines, wild geraniums) are blooming earlier than before.
- Edith's checkerspot butterflies are not migrating northward.

Global warming threatens the existence of currently endangered plant and animal species and puts others at risk as CO_2 levels rise. Global warming alters temperatures, humidity, soil, and vegetation; endangers natural habitat range; and can negatively affect other species in the ecosystem that endangered species depend upon for food.

There are several steps to extinction—it does not just happen overnight. Wildlife biologists have identified the key phases most species go through before they become extinct. The typical transition is from rare species to threatened to endangered to extinct species. The four phases are described as follows:

- Rare species: A plant or animal species whose population is small and isolated. Not many of its members can been seen in the wild, but its population is stable.
- Threatened or vulnerable species: A plant or animal species that may be abundant in some areas but still faces serious dangers. It is likely to become endangered in the near future.

- Endangered: A plant or animal species whose numbers have been reduced to such an extent that it is in immediate danger of becoming extinct. Such a species needs help from humans to survive.
- Extinct: A plant or animal species that no longer exists. No individual members can be found alive anywhere.

The key factor is to be able to identify these phases of extinction and catch and reverse the damage threatening species at the beginning. The closer a species gets to extinction, the more difficult it is to repair and reverse the damage.

Environmentalists have been warning that saving species is important because it promotes a healthy environment, of which humans are a part. Endangered species are nature's 911 call, because if they go, chances are the pollution and environmental degradation will ultimately kill us all.

PRESERVATION ISSUES

A January 2008 article in the *New York Times* stated that the efforts of conservation organizations whose goals are to preserve worldwide biologically rich ecosystems may not be enough to save species, in light of the effects of global warming and climate change. According to the article, some climate scientists believe efforts to reestablish or maintain salmon runs in the U.S. Pacific Northwest may not provide any long-term solution if global warming continues, making streams inhospitable. Another environmental preservation attempt that may prove to fail is that in the Florida Everglades. Current efforts to restore freshwater flow may be a waste of time if sea levels rise and submerge the area under ocean water in the future. In other preservation projects, geologists recommend abandoning efforts to preserve some of the fragile coastal barrier islands off the U.S. coast and focus more on allowing coastal marshes to migrate inland, buffering the effects of future sea-level rise.

With the documented steady upward climb of CO_2 in the atmosphere, both ecologists and conservation biologists are concerned that ecosystems may change so fast that many stable places will become vulnerable. According to Dr. Healy Hamilton, the director of the Center for Biodiversity Research and Information at the California Academy of Sciences, "We have over a 100-year investment nationally in a large suite of protected areas that may no longer protect the target ecosystems for which they were formed. New species will move in, and the target species will move out."

Because of this predicted evolution of changing climate and associated species relocation, conservationists today think they need to do more than just identify areas of acceptable current habitat and raise money to manage and protect those protected areas. Their thinking has evolved beyond that threshold to focus more toward the future. They must now make educated guesses as to which sites will be important 50, 60, or more years from now, and try to protect both species' ecosystems now and their needs in the future.

Bill Stanley, director of the Global Climate Change Initiative at the Nature Conservancy, says, "It's turning conservation on its head." The Nature Conservancy currently has plans to protect 10 percent of major habitat types, such as grasslands, freshwater environments, and forests, by 2015. Stanley continues, "We are not sure exactly how to treat this yet. Areas that we preserved as grasslands are becoming forests."

The issues center on whether there should be any human intervention on vegetation types and how to manage the resulting landscape. According to Dr. Hamilton, "Our whole strategy is going to have to shift."

Part of the dilemma is the unknown. Although conservation organizations believe the healthy, positive work they have completed on ecosystems to date has far-reaching positive effects and is highly valuable, the unknown centering on the rates and intensity of change makes it difficult to predict future outcomes and plan accordingly.

To illustrate this point, Dan Kimball, superintendent of the Everglades National Park, says that if sea levels rise faster than expected, mangroves could be submerged, which would have a negative impact on the health of the ecosystem. Conversely, if the rate of change is slow, mangroves could gather enough sediment and begin building a barrier to protect the terrain (which is what happened after Hurricane Wilma washed over a significant wetland in 2005).

THE IMPACTS OF LAND-USE AND LAND-COVER CHANGE

Another factor that scientists must take into consideration when analyzing the present and predicting the future effects of global warming is the influence that *land-use change* has on the land. According to the U.S. Environmental Protection Agency (EPA), a change in *land use* and cover can affect the Earth's temperature by changing how much solar radiation the land reflects and absorbs. This affects the Earth's energy balance. There are many ways that human action affects the landscape, such as agriculture, deforestation, reforestation, *desertification,* urbanization, and industrialization, which, in turn, change the Earth's energy balance. These modifications in land use can change the climate by altering atmospheric temperature, wind patterns, and precipitation distribution. Sometimes, the effects are more pronounced regionally.

According to the Union of Concerned Scientists, even small changes in urban development or deforestation can change local rainfall patterns and trigger other climate disruptions. In addition to altering the Earth's atmospheric temperature, land-use change can also affect the amount of carbon taken up and stored in vegetation (a process called sequestration) or released by the land surface.

Two of the oldest and most common transformations of land use by humans are the destruction of forests to create agricultural areas and the construction of cities in place of the natural environment. These changes can have a direct effect on regional climate systems. In cities, when natural ground cover is replaced by asphalt and other dark surfaces, the incoming solar radiation is absorbed by these dark surfaces and stored as heat, which it reradiates. Industry, homes, and cars also add significant sources of heat to urban areas, further upsetting a natural temperature balance. This phenomenon is referred to as the urban heat island effect and can influence climate on regional scales. In areas where forests have been cut down and vegetation cleared away for agriculture, the land heats up to higher temperatures than it did while it was still forest, also contributing to regional temperature rise.

When the balance of the landscape is affected, it will have an impact on the availability of water and the production of greenhouse gases. When forests and vegetation are cut down and cleared away, there will

Land use and ground cover help determine local temperature. Reflection from a city is much different than that from a natural area. In urban areas, industry, cars, buildings, and absorption of solar energy by dark surfaces (such as asphalt) create what is referred to as an "urban heat island." Natural wooded areas have a more moderate temperature range. *(Nature's Images)*

no longer be places to store CO_2, and long-term reservoirs of CO_2 will be lost. When areas becomes urbanized and highways are built, the burning of fossil fuels also adds a significant amount of CO_2 into the atmosphere, increasing global warming. Worldwide, natural areas are being transformed into urban areas and forests are being cut down. As the burning of fossil fuels becomes more and more of a problem, the negative effects of global warming are felt across bigger areas of the landscape. Each area of CO_2 input, whether from urbanization, defor-estation, industrialization, agricultural production, or other types of land use, contributes to global warming, increasingly a major issue the world's inhabitants face today.

Impacts to Forests

Forests have existed for hundreds of millions of years. Throughout the centuries, climate has forced forests to adapt by changing vegetation and migrating to new habitats. In the past, this has happened at a slow pace, allowing the adaptations to be successful. Today, changes in climate (temperature, precipitation, humidity, and air flow) are happening much more rapidly, and forests have not had the luxury of time to adapt.

At the same time, many other human influences have left their harsh imprint—clear-cutting, urbanization, deforestation, mining, and many other activities have destroyed large sections of forest, leaving them fragmented. According to the Intergovernmental Panel on Climate Change (IPCC), "It is in combination with these threats that the impacts of unprecedented rates of climate change can compromise forest resilience and distribution." According to the special Public Broadcasting Service (PBS) presentation "Earth on Edge," reported by Bill Moyers, nearly 9 percent of all known tree species are at some risk of extinction. Most of the world's forest loss has occurred in the last three decades—and most of this because of human activities.

Forests are extremely valuable ecosystems for many reasons. For example, they help to regulate rainfall and are also key sources of food and medicine. Forests provide abundant wildlife habitat, carbon storage, clean air and water, and recreational opportunities. They also contain a bounty of natural resources, such as wood, plants, berries, water, and wildlife. Forests also are of great aesthetic importance. The health and diversity of forests are largely influenced by climate.

Native forests have adapted to local climates. For example, in the far northern boreal forests, cold-tolerant species such as white spruce abound; in drier areas, conifer and hardwood forests thrive because they need less water. Since air temperature affects the physiological processes (such as growth) of individual plants, global warming has a strong influence over the health of forests.

This chapter explores the effects that global warming will have on the world's forests—temperate, boreal, and tropical—and the ecosystems that function within them. If proper management of forests does not take place now in preparation for what global warming will bring, forests may not survive.

TEMPERATE FORESTS

Temperate forests are those that lie within the temperate, or midlatitude areas on Earth, such as the United States. In terms of global warming and carbon dioxide (CO_2), forests play a crucial role. They store an enormous amount of CO_2: The trees, other vegetation, and soils within forests help to drive the global *carbon cycle* by sequestering CO_2 through photosynthesis.

As ecosystems, many forests are endangered today—mainly by human practices. Forests are being cut down so that the land can be used for other purposes, such as grazing land or agriculture. Other forests are being logged—hundreds of acres of trees are being harvested for wood products. Mining exploration and operations are encroaching into other forests. Urbanization and the construction of transportation networks to connect urban and industrialized areas have also claimed forested lands. Some forests have been deliberately set on fire.

Temperate forests are popular for recreational activities such as hiking, which these enthusiasts are doing on a warm summer day in the Uinta Range of the Rocky Mountains in northern Utah. *(Nature's Images)*

With all this activity occurring in the world's forests, this valuable resource is being lost at a rapid rate. Not only is carbon storage no longer available, but when trees are burned, all the stored carbon is released back into the atmosphere.

Carbon Sequestration

If forests are managed properly, they are one ecosystem that could be used to combat the effects of global warming. According to experts at the Union of Concerned Scientists (UCS), forests could reduce net carbon emissions by the equivalent of 10–20 percent of projected fossil fuel emissions through 2050. In the United States alone, measures to increase carbon storage in forests could sequester an additional 44–88 million tons (40–80 million metric tons) of carbon annually. The greatest potential carbon sequestration in forests is in the tropical and subtropical regions on Earth.

Carbon sequestration is viewed as a feasible way to offset global warming. According to the UCS, forest sequestration, also called forest-based mitigation, could occur via the following two strategies:

1. conservation of existing forests
2. increasing forests' carbon-absorbing capacity by (a) planting more trees, (b) promoting the natural regeneration of forests, or (c) improving forest management practices in order to increase *biomass*.

Once again, the UCS has stated that if forest sequestration is done correctly, it can slow climate change and the effects of global warming. For instance, based on their data, U.S. forests in the Pacific Northwest and Southeast could double the carbon they store if forest managers lengthened the time between harvests and allowed older trees to remain standing. Working with Canada and Mexico to attain the same goals would also help. Another suggestion proposed by the UCS is to create a system that allows private companies to get credit for reducing carbon when they purchase and set aside natural forests for conservation. There are abundant environmental benefits that come along with conservation, such as protecting biodiversity and maintaining clean watersheds.

Forest Values

One of the problems the UCS has identified is that the public must be better educated as to the valuable services that forests provide, such as biodiversity, carbon storage, and water purification. One of the major issues that must be faced today is the economics—when a quick sale can be made on the lumber or other goods taken from a forest, it is hard to convince people to turn away and choose to protect an environment that helps everyone but is intangible. The market-based versus environmental-based issues create some of the biggest challenges in dealing with global warming today.

Effects of Global Warming

Not all forests will have the same outcomes under the influence of global warming and changes in precipitation. Some forests will die back

Temperate deciduous forests have priceless aesthetic value, such as in the fall when the leaves turn to brilliant reds, oranges, and yellows. *(Nature's Images)*

(lose trees), while others may extend their ranges. Amounts of CO_2 will vary, adding its influence to the mix as well. Whatever the outcome of a particular forest, if its population cannot adapt or migrate with the changes, it will face extinction.

According to the IPCC, at least one-third of the world's remaining forests may be negatively affected by climate change during this century. Global warming may force plant and animal species to migrate or adapt faster than is physically possible, disrupting entire ecosystems. The IPCC also predicts that forests will have changes in fire intensity and frequency and increased susceptibility to insect damage and diseases.

According to the World Wildlife Fund (WWF), climate scientists use a variety of methods to predict the impacts of climate change on forests. On a global or large regional scale, they can predict shifts in ecosystems by combining biogeography models with atmospheric general circulation models (GCMs) that project changes in an environment where the CO_2 content has doubled. They also use biogeochemistry models to simulate the carbon cycle, flow of nutrients, and changes in precipitation, soil moisture, and temperature to study ecosystem productivity. They even have global models that simulate worldwide changes in vegetation composition and distribution.

The IPCC states that global warming both directly and indirectly affects forests during climate change. The direct impacts of warmer temperatures, changing rainfall patterns, and severe weather events can already be seen in certain tree and animal species. Even small changes can affect forest growth and survival—especially of the portions that lie along the outer edges of the ecosystem, where the conditions are marginal. When it gets hotter, more water is lost through evapotranspiration, which causes drier conditions and decreases plants' use of water. Water temperatures can also throw off the timing of flowering and fruiting for plants and adversely affect their growth rate. Forests will also be threatened when the seasonal precipitation patterns they have been used to change and water is not supplied when it is needed, causing drought conditions and stress, or supplies too much where it cannot be assimilated, causing flooding and mudslides.

According to Natural Resources Canada, the age and structure of a forest will also play an important role in determining how quickly it will respond to changes in moisture conditions. Mature (older) forests have well-established root systems, which means that they can tolerate drought better than younger forests or forests that have been disturbed in some other way, such as through disease infestation. Species type also plays a part—some species are more resistant than others.

Models have been run simulating the impact of elevated levels of CO_2 on forests. According to Natural Resources Canada, the results of these simulations are not as straightforward or conclusive as desired. Results are much more problematic. They found that higher levels of CO_2 improve the efficiency of water use by some plants, and some plants have been shown to adjust to higher CO_2 levels, but their absorption rates decrease over time. It is also hard to model CO_2 effects when other greenhouse gases may be involved, such as *ozone* (which works against CO_2) or nitrogen oxide (which can enhance tree growth).

In the short term (50–100 years), changes due to global warming will be focused on ecosystem function. In the long term, shift of forest types will be more significant. According to the IPCC, boreal (northern) forests will feel the most impact because their ecosystem will be greatly reduced since warming is expected to be most significant in the polar regions.

Some of the most vulnerable temperate forests will be the island, or isolated, forest communities, such as the fragmented forests encroached upon by urban and agricultural areas. There is nowhere for them to migrate. Forests located in high elevations on mountain systems face a similar threat—as they migrate upward, eventually there will be nowhere for them to go. Individual species that are indigenous to small geographic areas or have limited seed dispersal will also be threatened and endangered.

According to WWF, the following are the five major categories of changes expected in temperate forest ecosystems as a result of global warming:

1. Disturbance: Forest ecosystems will become more disturbed by extreme weather incidence and change in rainfall and temperature. Forests will become more fragmented and isolated, permanently altering the ecosystem.

2. Simplification: If global warming is severe enough, it will cause slower growing species to be replaced by faster-growing, short-lived weeds and other invasive species. This is a problem because it degrades the landscape, creating a species-poor forest—only a few invasive species dominate instead of the original rich, diverse mixture of many species.

3. Movement: Migration of species is expected both in latitude (toward the poles) and in elevation (up mountains). How fast individual species will be able to migrate is still unknown.

4. Age reduction: With all the stresses put on forest trees with global warming, the old growth stands are expected to die off, leaving younger stands in their place. This will have a negative impact on biodiversity.

5. Extinction: The most vulnerable forest habitats could disappear forever.

Under global warming conditions, surviving forests of the future will look very different from those of today. Although changes will vary in degree from area to area, all forests will be affected.

Climate Change Impacts

The forests in the United States and Canada are already showing the effects of global warming. Over the past century, a 1.67°–3.3°F (1°–2°C) increase in air temperature and changes in precipitation have already been documented. Experts believe that higher levels of CO_2 have already caused dieback of forested areas along the Pacific and Atlantic coasts.

It is expected that temperate forests in the United States will migrate northward from 62–329 miles (100–530 km) over the course of the next century. If the air temperature warms 3.3°F (2°C) over the same period, models have predicted that tree species will have to migrate 1–3 miles (1.6–4.8 km) per year, which is too fast for most temperate species, except for those whose seeds are carried by birds over greater distances. It is expected that grasslands will dominate many of these areas.

The wildlife inhabiting the temperate forests will also be affected by loss of habitat, drought, and wildfires. Diseases will also weaken them. According to the National Assessment Synthesis Team (NAST) of the U.S. Global Change Research Program, some forest models predict an overall increase in forest productivity with increased temperatures. They also clarify that other environmental factors, such as severe weather, could offset any productivity. NAST also states that a great deal of biodiversity change is expected in the United States. The maple/beech/birch forests in the Appalachian Range from New England to West Virginia could completely disappear. In the Upper Great Lakes region, aspen, birch, and red spruce could be obliterated.

If global warming occurs gradually, models suggest that oak/pine and oak/hickory forests will replace these threatened areas. If, however, conditions are bad enough, entire vegetation species could be lost.

NAST also found that warming in cooler areas—such as the northern United States and western mountains near the Canadian border—will increase tree richness and provide a better habitat for reptiles and amphibians. These same models also predict a decrease in bird and mammal richness in the eastern United States. NAST models have pre-

Temperate forests provide homes for many types of wildlife. If global warming causes forests to migrate or die off, it will threaten the existence of thousands of animal species that may not be able to adapt. *(National Park Service)*

dicted that temperate forests in Scandinavia will move northward and replace other forest communities. Models have predicted that China will see their coniferous forests reduced in size.

Recent Developments

Four examples of modern-day forested areas suffering the effects of global warming follow.

According to a July 7, 2006, report in *USA Today,* the frequency and size of large forest fires have increased dramatically in the last 20 years, and global warming is being blamed as part of the cause. The western region of the United States has seen an increase in wildfire occurrences since 1987; the wildfire season is now two and a half months longer than it was back then. Part of the problem is that spring warm-up is happening earlier and more quickly now than it has in the past. This sudden temperature transition melts the winter snow accumulation in the mountains, turning it into heavy spring runoff. Early green-up of grass, shrubs, and other vegetation promotes an abundance of biomass. When the summer months heat up significantly—there can be several days with temperatures in excess of 100°F (37.8°C)—and the winter melt has already run its course through the rivers and into reservoirs, the landscape becomes extremely arid in the hot, dry air, and the abundant biomass that grew during the ample spring runoff becomes extremely dry, posing a high fire hazard.

According to Anthony Westerling, a researcher at the Scripps Institution of Oceanography of the University of California, San Diego, "We didn't set out to make a climate change argument, but it's easy to see how a further rise in temperature under climate change would result in more frequent wildfires in these severe years. You get early snowmelt, the soil and vegetation dry out sooner, and you get a lot more fires, burning longer and getting bigger."

A group of scientists from the University of California, San Diego, and the University of Arizona studied 1,166 large forest fires that were each a minimum of 1,000 acres (405 ha) in size from 1970 to 2004 on national forest and parklands. They determined that in the latter part, from 1987 to 2004, there were four times as many forest fires and more than six times as much total land area burned. The fires in the latter part were also more difficult to control and lasted longer. The duration of fires began averaging five weeks. The most famous fire that occurred during this time was the Yellowstone fire in 1988. One-third of Yellowstone National Park burned, and the fire was not extinguished until it finally burned itself out in the fall.

Tom Swetnam at Arizona's Laboratory of Tree-Ring Research says that climate is the principal factor causing large wildfires, especially in the northern Rockies. He backs up his conclusion by pointing out that the number of large fires in Wyoming, Idaho, and Montana has risen 60 percent since 1987. The combination of rising temperatures and drying is making forests highly susceptible to wildfires.

Tourism and Recreation

One recreational activity that may be hard-hit by the effects of global warming is the ski industry. Already, many of the glaciers in the Alps in Europe are melting, threatening the future of ski resorts. Because of

The wildfire that burned one-third of Yellowstone National Park in 1988 was one of the most destructive fires of all time. *(National Park Service)*

already warmer conditions and changed weather patterns, many winter recreational resorts have to use snow machines to make artificial snow. In an article in the *New York Times,* Shardur Agrawala, a climate specialist, noted that alpine resorts, especially those at low altitudes, are developing other revenue opportunities that do not require snow, such as spas, restaurants, shopping centers, swimming pools, convention centers, and tram rides geared for the entertainment of hikers and sightseers. According to the same article, some resorts in Europe claim global warming has already begun affecting tourism. Roman Codina, the managing director of a mountain lodge in Zermatt, Switzerland, testified about a surge of visitors during the summer months—tourists from the hotter European cities seeking relief in the cooler mountains.

Ski resorts in the United States stand to face a similar dilemma. Already, resorts create artificial snow to establish a snow base in years when there is not an adequate accumulation to open the ski resorts by Thanksgiving in November—a traditional start of the ski season. Artificial snow is also used as a backup for ski competitions during drought-like years.

BIRDS, GARDENS, AND GLOBAL WARMING

According to a June 3, 2003, article in National Geographic News, mild winter temperatures that occurred in 2002 in Britain, along with an unusual surge in the rodent population, caused an overwhelming amount of birds to be born. The increasing temperatures of global warming were said to be the cause. David Glue, a research biologist with the British Trust for Ornithology (BTO), said "We had five cases of tawny owls laying eggs from Christmas week right through to the end of January. This is exceptional—usually they don't start nesting until the second or third week in March."

Another factor that he said contributed significantly to the amount of birds being born was the large availability of rodents as a food source. He claimed that 2002 went down in the records as the fourth hottest year since official records began being kept in 1659. He also confirmed that because the birds nested so early in the year, many of them went on to hatch another brood later on that same year, some as late as December, which was highly unusual.

In the past, winter nesting has been rare in Britain—the climate has been simply too cold. But David Glue said, "We can expect earlier and earlier egg laying and cases of birds nesting right through the winter. This is a reflection of climate change through global warming. Usually it's increasing day length that triggers egg laying, but if you get exceptional spells of warmth, along with an unseasonal super-abundance of food, this can override day length as the main factor."

Glue was also able to support these conclusions with BTO's records kept in the Nest Record Scheme, which represents the largest database of nest histories in the world (more than 1.2 million individual nest histories for British birds since 1939). According to Humphrey Crick, BTO's senior ecologist, "About a third of the 60-odd species we've looked at show statistically significant trends toward earlier egg laying. This is strong information because it comes from all over the country and isn't restricted to just one

Spread of Insects

According to a report in National Geographic News on November 5, 2001, a specific species of mosquito, *Wyeomyia smithii*, has gone through an evolution and been able to adapt to the climate changes

study site." He also said that birds are laying eggs between one and three weeks earlier than they did 15 years ago. The problem with some species is that their hatch times might not be in sync with the availability window of the particular insects they like to eat. If the life cycle and availability of a food source is not timed to their needs, it puts the birds at risk of starvation. BTO scientists have also noted that while some bird species are currently migrating to Britain from southern latitudes, such as egrets, European bee-eaters, and Mediterranean gulls, breeds common to Britain, such as dotterel and snow bunting, are leaving Britain and migrating farther north.

Another endangered ecosystem component in Britain is their private and public gardens. According to the United Kingdom Climate Impacts Programme (UKCIP), the drier summers and milder winters that are seeming to dominate the usually cool and rainy climate in Britain, are beginning to hurt Britain's renowned verdant green lawns, flowering shrubs, plush rose bushes, and thick clustering vines that draw visitors from around the world. The unwelcome change is being blamed on global warming.

The beautiful gardens exist in Britain for a good reason—gardening is their leading hobby. In fact, 41 percent of the population (about 27 million people) engage in some type of gardening. Because of current and projected changes in climate, hotter and drier summers are expected. This means that spring flowers will bloom earlier, winters will be milder with less snowfall. This will increase rainfall and cause flooding. Gardens are migrating south 2.5–4.5 miles (4–7 km) a year.

Summer droughts are also having a negative impact on gardens. Chris West, a scientist at UKCIP, said, "Some climate change is now inevitable and although we can still influence the extent of this for the latter part of this century, the die is already cast for the next 50 years. We want people to find out more about how these changes will affect their lives and to consider this in their plans for the future."

caused by global warming. William Bradshaw, an evolutionary biologist at the University of Oregon, has found that this tolerance can occur in as little as five years' time.

Because winters are gradually getting warmer, the mosquitoes are now breeding later in the year. Compared with their hibernation habits 30 years ago, northern populations of the mosquito have adapted to milder winters and become dormant later in their annual cycle. In fact, hibernation today is about nine days later than it was in 1972.

The problem with this altered breeding and hibernation schedule is that it has an impact on the natural food chain and ecological system. If the timing of the component in a system gets out of synch, then food sources may not be available when predators need them, upsetting the entire balance of the ecosystem.

Wildlife

One noteworthy contribution to atmospheric methane—a serious global warming gas in the atmosphere—is, surprisingly, from livestock. In a report May 13, 2002, on National Geographic News, scientists in New Zealand determined that livestock in their country were adding to the global warming problem.

Katherine Hayhoe, an atmospheric scientist at the University of Illinois, remarks, "New Zealand is unique in that more than 50 percent of its greenhouse gas emissions arise from methane released by enteric fermentation."

Enteric fermentation is methane that is produced as part of the digestive process of cows, sheep, and other animals. It is released when the animals burp, which is, surprisingly, a lot. There are roughly 10 million cattle and 45 million sheep in New Zealand, and their collective burps add up to 90 percent of New Zealand's methane emissions. According to the U.S. Environmental Protection Agency (EPA), livestock in the United States only account for approximately 2 percent of methane emissions.

To combat this problem, scientists in New Zealand have discovered that by feeding livestock plants that are high in *condensed* tannins they produce up to 16 percent less methane. Tannins are naturally occurring plant polyphenols that bind and precipitate proteins. They are found in legume forages and some grasses. Scientists view this as a significant step to reduce the effect livestock are having on the global warming problem.

According to Michael Tavendale, the researcher who made the discovery, "It's early, but this is very encouraging news that will give our research new impetus and offer positive opportunities to New Zealand farmers for controlling the problem."

This diet has other benefits as well: It increases livestock weight gain, improves milk yields, and decreases parasites. As a comparison, a 200-cow dairy herd has the "petrol equivalent" of 6,400 gallons

KATRINA AND THE CARBON BALANCE

When Hurricane Katrina roared through New Orleans and the Gulf Coast in 2005, its destruction extended not only to manmade structures but to the natural surroundings as well. According to a *New York Times* report issued on November 20, 2007, entitled "Katrina's Damage to Trees May Alter Carbon Balance," the world's most destructive hurricane also uprooted or severely damaged 320 million trees as it cut its disastrous path across the Gulf Coast.

Dr. Jeffrey Q. Chambers of Tulane University conducted a study in November 2007 using satellite imagery before and after the storm to determine the change in deadwood and ground litter. This is important because when a tree is alive, it acts as a storage reservoir for carbon. Conversely, when vegetation dies, sequestering of carbon stops. Once vegetation dies, the process of photosynthesis stops, the organic matter begins to decompose, and the carbon stored within it is released into the air. The more carbon released, the worse for the problem of global warming.

When asked about forest regrowth, Dr. Chambers replied, "It takes a lot longer to recover the biomass than it does to lose it. In some cases, it is possible for a forest to go from being a net storer of carbon to a net source."

A very serious problem for the future is that as global warming increases, more severe weather events will occur, leading to damage to trees and the resulting addition of carbon to the atmosphere, a vicious cycle.

(24,000 l) of gasoline, the gas necessary to drive an average car 124,274 miles (200,000 km).

Controversial Findings

A study conducted by the Carnegie Institution in Washington, D.C., and the Lawrence Livermore National Laboratory in California led their scientists to conclude that forests in temperate regions actually could worsen global warming. Using complex modeling software to simulate changes in forest cover and their effects on global climate, the results they obtained from their model were surprising.

Ken Caldeira from the Carnegie Institution remarked, "We were hoping to find that growing forests in the United States would help slow global warming. But if we are not careful, growing forests could make global warming even worse."

The researchers concluded from their study that it was the tropical forests that cooled the environment because they released great quantities of water during the evapotranspiration process. Forests in the temperate regions, however, not only did not release as much water, but also absorbed a lot of sunlight, heating up the Earth. To illustrate this point, during one of the computer simulations, they covered the Earth with forests north of the 20-degree latitude line. They found that the surface air temperature responded by rising more than 6°F (3.6°C). In another iteration, covering the entire planet's land area with trees increased temperatures only about 2°F (1.2°C).

The warming was not evident at first. During the initial phase, there was cooling because the trees took up carbon dioxide, which offset the absorption of sunlight. But this effect did not last. The absorption of energy lasts forever, but as trees mature, they sequester less carbon. Based on this simulation, the involved scientists concluded that planting a forest in the United States could cool the Earth for a few decades but would lead to planetary warming in the long term. The scientists determined that a global replacement of current vegetation by trees would lead to a global warming of 2.4°F (1.4°C). If areas were replaced with grasslands, it could lead to a cooling of 0.7°F (0.4°C). This situation does not hold true for tropical forests, however. Rain forests keep the Earth cool not only by absorbing CO_2 but also by evaporating water.

BOREAL FORESTS

The effects of global warming are felt most at the poles, where it is predicted that temperatures could climb 8.3°–17°F (5°–10°C) or higher over the next century. It is estimated that warming will have a negative effect on the species that live in the ecosystems and that approximately 24–40 percent of the species living in the boreal forests right now will be lost.

Simply put, species that live in the north will be crowded out by species from the temperate regions that will be migrating northward in search of cooler climates. The species that will invade the present boreal forests will be today's temperate forest species and grasslands. According to the IPCC, as the boreal forest vegetation is forced out, it will migrate poleward 186–311 miles (300–500 km) in the next century. Proof of this can already be seen in western Canada; plant zones there have already begun shifting poleward.

There are major challenges this migrating vegetation will encounter. To begin with, the soils in the tundra region are not fertile and conducive to high-density vegetation or tree growth. According to the IPCC, they lack the biota necessary for colonization. Specific seed dispersal rate and migration tolerance range are also important factors that might keep trees from being able to survive the poleward migration rate set up by global warming. As an example, white spruce can colonize 62–124 miles (100–200 km) over 100 years, and Scots pine can migrate 2.5–5 miles (4–8 km) every 100 years. According to A. Solomon and K. Jardine of Greenpeace, the rate of global warming will be about 10 times faster than what is needed for successful species migration at an average migration rate of 16 miles (25 km) each century through natural seed dispersal.

There are other factors that will hurt species migration as well, such as habitat fragmentation (small isolated clusters instead of one large cohesive unit) and competition from more hardy species. As temperatures change, it may also affect the timing and rate of seed production, which will affect the growth and strength of trees. Trees that have limited seed dispersal mechanisms will also suffer, i.e., trees whose seeds are carried long distances by the wind will have a better chance of survival than those whose seeds fall closer to the tree.

The ability to adjust to a larger temperature range will also play an important role. Vegetation with narrow temperature tolerances will be vulnerable to extinction. The IPCC has stated that a drastic change in species composition and loss of habitat with even a 3.3°F (2°C) warming near the poles will damage the ability of an ecosystem to function as species richness begins to be killed off. They warn that a decrease in habitat if this were to occur would result in the loss of 10–50 percent of all the animals living in the boreal Great Basin mountain ranges.

According to Natural Resources Canada, an average rise in temperature of 1.7°F (1°C) over Canada in the last century has had a negative impact on vegetation. At mid to high latitudes (45° N–70° N) plant growth and the length of the growing season has increased. In portions

The northern boreal forests provide habitat for animals like the caribou. Global warming will cause boreal forests to shrink in size as other ecosystems move northward to cooler climates, crowding them out. *(National Park Service)*

of western Canada, there has been a decrease in rainfall as temperatures have risen, and this has hurt the growth of some tree species, such as aspen poplar. In Alberta, aspen are now blooming 26 days earlier than they were 100 years ago.

Another major concern for boreal forests in a warmer climate is insect infestations. Insects commonly found in temperate forests, such as mountain pine beetle, will migrate north along with the forests and continue to infect as they move northward.

As temperatures climb, droughtlike conditions may develop. If this happens, there will also be greater incidences of wildfire. In the past 40 years, the trend has already been established that as climate warms, wildfires have become more frequent and are burning larger areas. This overall trend has been seen in places such as southern California and also in boreal locations in Canada and Russia. As global warming continues, longer fire seasons, drier conditions, and more frequent severe electrical storms are projected to increase, causing fire seasons to become more problematic as the climate continues to change.

While it is true that some forest species' seeds are actually dispersed by fire, which will aid their migration, and burned litter will add nutrients to the soil, over time reoccurrences of wildfire will fragment established vegetation colonies and make it more difficult for them to migrate. In addition, as older trees burn, they will add carbon to the atmosphere, and as younger trees replace these burned areas, there will be less initial carbon storage capability.

TROPICAL FORESTS

The world's tropical forest ecosystems are very sensitive to disturbances such as overgrazing, logging, plowing, and burning. Converting natural ecosystems to agricultural and logging uses combined with global warming pose the largest threat to rain forests today. According to the Food and Agriculture Organization of the United Nations, a land area the size of Ireland or South Carolina is lost to these uses every few years in the rain forests. Developing countries are hit harder than developed countries. Unfortunately, it is often seen as necessary to farm the land in pursuit of food or to sell the land to logging companies in exchange for needed income.

This rain forest in Malaysia is full of lush vegetation. Because the root system does not extend deep underground, the tall trees have an external buttress root system to help prop them up. *(Rhett Butler, Mongabay.com)*

Vegetation Migration and Deforestation

According to José Antonio Marengo of Brazil's National Institute for Space Research, if the issue of global warming is not addressed today, it will have a negative impact on the rain forests. They will receive less rainfall and experience higher temperatures, enough so that it could transform the Amazon—the world's largest remaining tropical rain forest—into a savanna (grassland) by the end of the century. The Amazon covers almost 60 percent of Brazil and hosts one-fifth of the world's freshwater and nearly 30 percent of the world's plant and animal species. José Marengo is involved in research focused on the following two scenarios:

1. If no action is taken to slow or halt global warming or deforestation, temperatures will rise 8.3–13°F (5–8°C) by 2100,

and rainfall will decrease by 15–20 percent. This is the scenario that would transform the Amazon into a savanna.

2. If action is taken now to slow global warming, temperatures would most likely rise 5–8.3°F (3–5°C) by 2100, and rainfall will decrease by 5–15 percent.

Marengo stresses that "if pollution is controlled and deforestation reduced, the temperature would rise by about 8.3°F (5°C) by 2100. Within this scenario, the rain forest will not come to the point of total collapse." Marengo also stated that he was optimistic that the worst-case scenario could be averted, but that it would require a major effort by industrialized nations to reduce emissions of greenhouse gases, like CO_2.

Two researchers are currently involved in detecting early drought conditions in the rain forest through satellite technology. Dan Nepstad,

The rain forests provide homes to abundant amounts of wildlife. In Malaysia, this long-tailed macaque swings high up in the trees. *(Rhett Butler, Mongabay.com)*

an ecologist at the Woods Hole Research Center, and Greg Asner of the Carnegie Institution and Stanford University are working in an area of the Amazon conducting simulated drought experiments.

Dan Nepstad says that the southeastern part of the rain forest actually goes for months at a time with little or no rainwater and the trees survive by sending deep taproots 66 feet (20 m) down into the ground to trap water. While severe El Niño events can cause drought-like events, Nepstad and Asner are most concerned that global warming will bring longer and more severe droughts to the rain forests. In addition to El Niño, they also testify that large-scale deforestation and burning biomass jeopardize cloud formation and rainfall, contributing to droughtlike conditions.

Because of this, their goal is to turn to space for the answer. Through the use of satellite imagery, they are conducting research to be able to identify the precise signals of a drought-stressed forest. If satellite imagery can detect signs of stress in the vegetation well before human eyes can detect it, it would give a critical advantage to farmers, ranchers, timber operators, fire managers, land managers, and conservationists. Positive action could be taken before significant damage was done and land degradation became extreme.

In response to their efforts, Brazil's Large-Scale Biosphere-Atmosphere Experiment in Amazonia (LBA), the largest cooperative international scientific project ever to study the interaction between the Amazon forest and the atmosphere and climate, was created in 1996. Their work today focuses on how to use imagery to monitor forest ecosystems, drought, and the increasing effects of global warming.

Unfortunately, many of the world's rain forests are being cut down at an accelerated rate for agriculture, pasture, mining, and logging. When these forests are cut down or burned, huge amounts of CO_2 are reintroduced into the atmosphere. Tropical deforestation accounts for about 20 percent of the human-caused CO_2 emissions each year to the Earth's atmosphere. This makes rain forest deforestation an important issue when dealing with the challenges of global warming.

The rain forest is full of plant species—many not even studied yet. Many medicines used today to cure serious illnesses come from the rain forest. Scientists who research these dense, tropical areas use suspended bridges as catwalks, such as this one in Malaysia, strung from tree canopy to canopy. *(Rhett Butler, Mongabay.com)*

Deforestation, shown here in Madagascar, is a huge problem and has become so widespread that its connection to global warming cannot be overlooked. These huge resultant erosional features are called lavaka. The world's rain forests must be managed properly to avoid this kind of environmental damage. *(Rhett Butler, Mongabay.com)*

Effects on Wildlife

In the rain forest, some of the species being negatively affected by global warming are parrots. The black-cheeked lovebird (*Agapornia nigrigenis*), an African parrot species found in Zambia, is one parrot species already feeling the heat of global warming. Only located in a small area in Zambia, there is only one wet season—from November through April; the rest of the year is dry. The lovebird's survival depends on the availability of fresh, freestanding water. Water in the area is already a scarcity, and, as global warming continues to increase, the lovebirds may not be able to adapt. In general, the smaller the geographic habitat and the more specialized the niche of a species, the less likely it is that it will be able to adapt to climate change.

In Australia, there are three avian species that are currently vulnerable under global warming conditions. The endangered golden-shouldered parrot (*Psephotus chrysopterygius*) is being threatened by wildfires. The orange-bellied parrot (*Neophema chysogaster*), which has a population of only 200 individuals in coastal areas, is being threatened by the flooding of coastal salt marshes as sea levels rise. The budgerigar (*Melopsittacus undulates*), or "budgie," is having trouble breeding as Australia experiences its worst drought in more than a century. Tens of millions of budgies have died since the drought intensified in 2003.

In South America, green-rumped parrotlets (*Forpus passerinus*) have suffered from changes in rainfall. Both too much and too little rain discourages nesting. In periods of drought during La Niña events in Patagonia, the surviving chicks of the burrowing parrot (*Cyanoliseus patagonus*) had a higher incidence of dying; those that survived had

There are several layers of canopy in the rain forest, and all are teeming with wildlife, such as this toucan in the Tambopata National Reserve in Peru. *(Rhett Butler, Mongabay.com)*

Budgies are dying in Australia due to drought. Tens of millions have perished in the wild since 2003. Budgies kept in captivity in the United States are commonly referred to as parakeets and are popular for their beautiful songs and chirps. *(Nature's Images)*

stunted growth. In Costa Rica, there have been upward shifts in ranges of mountain populations.

In the rain forests, just a 1.7°F (1°C) increase in temperature affects egg-laying dates, leaf and fruit emergence, disease patterns, and river flow. Steve Boyes, an African parrot researcher, warns that global warming is causing an ecological imbalance that species are reacting to and not evolving with. In the past, parrots have timed their breeding with the cycles of high-protein food sources, but since global warming could change fruiting and flowering times, breeding seasons could shift to times of the year when nesting behavior is not as well suited for the parrots.

As a group of wildlife species, parrots are the most endangered order of birds. Currently, 31 percent of rain forest species are at risk of extinction. According to an article in *BirdTalk* magazine, the International Union for Conservation of Nature (IUCN) is currently assessing the impact of climate change on species and trying to understand how to deal with species at risk by putting them on the IUCN Red List of Threatened Species. The IUCN stresses the importance of taking action now, even though effects in some cases may not be felt on the Earth's ecosystems for years to come.

ADAPTATION

In view of the impacts that may be felt by the world's forests—temperate, boreal, and rain forest—it is important to begin looking at adaptation strategies today. The WWF has proposed the following as viable adaptation options in order to reduce the threat on forest ecosystems from global warming and climate change:

- Reduce present threats that could harm the ecosystem, such as degradation and the introduction of invasive species.
- Manage large areas of land in a comprehensive way tied to the landscape. Focus on all the components that compose a large-scale area and plan for the adaptation migration of different species; develop special plans for sensitive, vulnerable areas that require greater attention; and plan for potential habitat needs in the future.
- Provide buffer zones and flexibility of land uses. As species migrate in response to global warming, the land must be managed to accommodate those movements. Buffer zones will help protect lands bordering reserves as conditions inside the reserves become uninhabitable, causing a species to migrate.
- Protect mature forest stands. Mature trees are better able to withstand large-scale environmental changes and can provide a safe habitat for other species to adapt within.
- Maintain the natural fire regime for the area. Different ecosystems have different fire ecologies; some forests that are not allowed to burn suffer from lower diversity; others that are allowed to burn destroy biodiversity. Because of this, a fire management plan needs to be specifically tailored for the area it encompasses. Properly developed plans reduce the intensity and spread of wildfire and its impacts.
- Pests must be actively managed. Because global warming is associated with invasions of insects, disease, and exotic species, healthy management practices must be set in place to protect the forests. Prescribed burning and the use of nonchemical pesticides are two methods of controlling insect infestations. According to research done by Natural

Resources Canada, *baculoviruses* could also be used to attack pest species, such as spruce budworm, while leaving the rest of the environment alone.

Different geographical forest areas will require different monitoring, planning, and protection strategies. But no matter what type of forest it is, or where it is located geographically, it plays an important part in the Earth's natural response to climate, and it is imperative that humans understand the unique ties they have with the forests and their responsibilities as land stewards.

Impacts to Rangelands, Grasslands, and Prairies

Native grasslands cover about one-quarter of the world's land surface, making them a significant component of the world's vegetation. Because of this, changes brought upon them by global warming could have a serious impact on a major global ecosystem. The extent and health of natural grasslands are typically controlled by rainfall and fire. These two factors limit the extent of their range, and, with an increase in global warming, it is already known that there will be a decrease in rainfall in some areas and an increase in wildfires.

The world's grasslands will face serious consequences from global warming. These regions provide grains and crops for people, rangeland for cattle and sheep, habitat for wildlife, such as wild horses, buffalo, deer, elk, giraffes, zebras, kangaroos, leopards, and elephants, and play a role in carbon sequestration and maintaining an overall healthy ecological balance in nature. Without grasslands, ecosystems would be imbalanced with negative results. This chapter will examine the importance of grasslands and their role in the world's ecological health and what global warming will do to them as the Earth's atmosphere continues to heat up. It will look at some ecological anomalies that scientists have found in some of the world's grasslands. This chapter will then focus on some specific grasslands around the world and explore some of the proposed options for adaptation in a warming world. Because grasslands are an important ecological zone, understanding their significance and taking action now will help reduce disaster in the future.

THE IMPORTANCE OF GRASSLANDS

The world's grasslands are dominated by grasses and *forbs* and have few trees. The dominant wildlife that live in this ecosystem are grazing and roaming animals. Grasslands can be either temperate or tropical. The Great Plains of North America is an example of temperate grasslands. Grasslands are different than cultivated agricultural lands. Grasslands have a great diversity of species and have evolved and developed naturally over millions of years, affected by climate. Cultivated lands focus on the growing of a few select species—such as wheat—lowering their biologic diversity.

Grasslands are also important for carbon storage. Whereas carbon is stored in trees, plants, and organic matter on the surface of the soil in forests, in grasslands it is stored in the soil. Due to human development, most of the Earth's grasslands suitable for agriculture have already been developed. According to the Food and Agriculture Organization (FAO) of the United Nations, as much as 70 percent of the Earth's 3.2 billion acres (1.3 billion ha) of grasslands has become degraded, mostly because of overgrazing.

Temperate grasslands are located north of the tropic of Cancer (23.5° North latitude) and south of the tropic of Capricorn (23.5° South latitude). The four major temperate grasslands (which go by different names depending on where they are located) are the plains of North America, the veldts of Africa, the steppes of Eurasia, and the pampas of South America. These areas are dominated by grasses, with a few trees, which are limited by drought, wildfire, and grazing.

The principal places that trees, such as cottonwood, oak, and willow, are found are along river valleys because they need a reliable water source nearby. Flowers are commonly found among the grasses. The most common grass types are buffalo grass, purple needle grass, blue grama, and galleta. The soil in temperate grasslands is rich in nutrients because of the growth and decay of deep, dense grass roots. The roots serve to hold the soil together as a cohesive unit. In fact, some of the world's most fertile soils are found in the grasslands of the eastern prairies in the United States, the steppes of Russia and Ukraine, and the pampas of South America.

Tropical grasslands are located near the Earth's equator, between the tropics of Cancer and Capricorn. These are the grasslands of Africa, Australia, India, and South America. Tall grasses dominate these landscapes—some as tall as 3–6 feet (1–2 m). These are found in tropical wet and dry climates, hot all year long, their temperatures never falling below 64°F (18°C). Usually dry, there is a season of heavy rain, bringing annual rainfall to 20–50 inches (51–127 cm) per year. One of the biggest threats to the tropical grasslands from global warming is desertification.

According to the Intergovernmental Panel on Climate Change (IPCC), the structure and function of the world's grasslands make them one of the most vulnerable to global climate change of any of the terrestrial ecosystems. Grasslands are naturally vulnerable to encroachment by nonnative, invasive species (such as cheatgrass). As temperatures rise and precipitation decreases due to global warming, grasslands will become stressed under droughtlike conditions, causing them to become more vulnerable to wildfires. Once affected by wildfire, grasslands face additional ecosystem and habitat destruction. Following wildfires, native vegetation often does not grow back because it must compete with more aggressive, robust invasive species. In the Southwest, for example, it is a common problem for cheatgrass to take over during the regrowth phase that occurs immediately after a wildfire. In addition, global warming will also cause a gradual decline in soil carbon and nitrogen content, having a negative affect on the health of the returning vegetation.

According to an article in ScienceDaily (1/15/01), scientists have experimented to determine how much of a carbon sink grasslands have the potential to be. Dr. Shuijin Hu, an assistant professor of plant pathology at North Carolina State University was involved in a study that showed grasslands are able to act as carbon sinks when carbon dioxide (CO_2) concentrations in the atmosphere are increased. In his study, he discovered that grasslands actually sequester carbon in the soil. Soil microbes respond to the changes in carbon and nitrogen and actually store excess carbon as atmospheric levels increase. Sequestration occurs when the amount of CO_2 absorbed by vegetation is greater than the amount of gas released by decomposing plant material.

In the study, Dr. Hu, along with Dr. F. Stuart Chapin III of the University of Alaska, Dr. Christopher B. Field of the Carnegie Institution of Washington, D.C., and Dr. Harold A. Mooney of Stanford University, focused for five years on a study site at Stanford University's Jasper Ridge Biological Preserve in California. From 1992 to 1997, two study plots were maintained—one at today's normal CO_2 levels (360 ppm) and the other at double that amount (720 ppm). Hu explained, "Other scientists have proposed that grasslands can act as carbon sinks when atmospheric carbon dioxide is elevated." The research they conducted showed that grassland soils can sequester carbon and also found a trend toward increased soil carbon under elevated CO_2 conditions.

At the end of the experiment, they found that the soil core sample from the plot that had received double the normal CO_2 had stored more carbon. Their results presented an interesting relationship in grasslands. When CO_2 is increased in the atmosphere, grassland plants grow more rapidly, while drawing nitrogen from the soil. This nitrogen depletion ends up keeping it from the soil microbes, which in turn reduces their ability to decompose dead plant material (which releases CO_2 to the atmosphere). There does seem to be a tipping point, however, where only so much carbon can build up. Once it reaches that level, the lack of nitrogen keeps additional plants from growing. Dr. Hu concludes that "Forests may be of greater potential as a long-term carbon sink than annual grasslands because trees can sequester carbon in above-ground biomass." Grasslands do, however, play a role in carbon storage.

IMPACTS OF GLOBAL WARMING

According to the IPCC, general circulation models (GCMs) predict grassland ecosystems will experience climatic changes such as higher maximum (daytime) and minimum (nighttime) temperatures and more intensive precipitation events. In fact, recent studies of global warming processes reveal a surprising situation—daytime high temperatures are not the problem; it is the nighttime highs that are causing problems. During late winter and early spring, the situation is most pronounced. Nighttime high temperatures have risen, changing the temperature regime enough that the last frost date occurs on average two weeks earlier now than it did 20 years ago. In addition, native grasses are now germinating later. Invasive and noxious plants that have invaded grasslands over the years are germinating early, taking the moisture out of the soil and using the nutrients that would have been available for the grasses. In addition, cattle grazing occurs on much of the grassland areas, and they do not eat the weeds, compounding the problem with invasive species.

Based on the results of the IPCC studies, the world's grasslands that are expected to receive an increase in precipitation, such as those in the western United States, may not be faced with drought from lack of water and rising temperatures, but may suffer from accelerated nutrient cycling, which in turn could encourage the spread of more invasive species. In the hot desert grasslands (such as in Australia and the Sonoran and Chihuahua Deserts of Mexico and the United States), GCMs predict an increase in the frequency of intense rainfall and flash floods. These areas are expected to have increased erosion and nutrient loss. Thus, ecosystems that have already been disturbed in other ways, such as by wildfire or desertification, will be affected even more and have a more difficult time surviving under global warming.

In a study conducted on May 30, 2007, by the Agricultural Research Service (ARS) and Colorado State University, it was determined that global warming may lower grassland quality. Based on their findings, rising CO_2 in the atmosphere and temperatures may initially lead to greater plant production, but it will eventually lead to a decline in soil carbon and nitrogen. This would present a serious problem to ranchers because grazing cattle and sheep require nitrogen-rich vegetation for their digestive processes. The study determined that if grassland quality was negatively affected, it would not only harm livestock but also the native animals that have grazed there for centuries.

Buffalo once roamed the vast grasslands of the United States. Buffalo are beginning to make a comeback in some places today, but if global warming threatens the health of grasslands, their habitat may be destroyed. *(National Park Service)*

There are many threats to grasslands from climate change. Continued global warming could convert marginal grasslands into deserts as rainfall decreases. The threat of desertification is all too real in many parts of the world. As areas struggle to find water to support populations, cultivate crops, and support livestock, grassland ecosystems will face further degradation.

SOME SURPRISES

One surprising two-year study conducted by Stanford University and the Carnegie Institution showed that in some cases wetter grassland ecosystems could also exist as a result of global warming. The model they developed ultimately contradicted other climate models that predicted drier grassland landscapes. In this particular experiment, instead of the soil drying up, higher temperatures actually increased soil moisture by 10 percent. According to Erika Zavaleta of the University of California, Santa Cruz, "We were surprised to find that warming actually increased moisture in our grassland plots during those critical weeks in late spring at the end of the growing season when moisture shapes which plant species prevail. We traced this unexpected moisture increase to the plants themselves." She added, "This doesn't mean climate change is good for California's grasslands, but it reinforces the importance of paying attention to how plants and animals could modify its effects."

What the researchers concluded was that when the plants shut down, the moisture becomes trapped in the soil. During the two-year experiment, excess heat and CO_2 were applied to their grassland test plots. Researchers were surprised to discover that moisture increased in every plot where the heat lamp was kept on and soil moisture increased in plots that had been exposed to increased levels of CO_2.

"Simulated warming increased spring soil moisture by five to 10 percent under both ambient and elevated CO_2," the researchers said.

One of the natural treasures of the American West is the free-roaming herds of wild mustangs. Finding food and water can be a challenge, and during years of drought herds can be hit especially hard. These wild herds will be hit severely and their natural habitat threatened if global warming intensifies. *(Kelly Rigby, Bureau of Land Management)*

"While elevated CO_2 has been shown to increase soil moisture in other field experiments, stimulation of soil moisture by warming has not been previously reported."

"The study adds to a growing body of knowledge about the major role that plants can play in global warming," said coauthor Christopher Field, a professor, by courtesy, of biological sciences at Stanford and director of the Carnegie Institution's Stanford-based Department of Global Ecology.

The result from this experiment was not what the scientists expected. Their expectations were that higher temperatures would increase the amount of evapotranspiration—water evaporation from the soil and the surface of the plants. At the study site, they determined that soil moisture evaporates through the plants. Of extreme significance was that the warming initially caused the early death of many of the grasses and wildflowers. Some of the plots lost 17 percent of their initial green vegetation. The researchers believe that the death of these early grasses explains the unexpected rise in soil moisture. According to Dr. Zavaleta, "Simulated global warming accelerated the death of the dominant grasses in our plots, leaving slightly more water in the soil for other species later."

Scientists take the anomalies from this discovery and are able to apply them to other areas worldwide. In particular, they believe that they could apply the concepts learned at the test site to Mediterranean-type ecosystems because they experience similar growing season dynamics. This opens the possibility that these select areas may respond in the same way to global warming, necessitating their long-term management response to be tailored for a different outcome—one of increased moisture content rate than decreased.

GRASSLANDS AROUND THE WORLD

According to a news release issued by the French National Institute for Agricultural Research (INRA), the European GreenGrass project, a study conducted in Europe by Jean-François Soussana, determined that the least productive sites—those closest to mountains or at higher latitudes—stored less carbon than the grasses in the more productive areas located in the plains. The more productive sites, however, were grazed more and emitted more methane and *nitrous oxides,* making them more vulnerable to the effects of global warming. They determined that the grasslands they studied served as a moderate carbon sink.

The extreme heat wave Europe experienced during the summer of 2003 gave scientists a rare opportunity to study its effects and take measurements. That summer, temperatures climbed 10°F (6°C) higher than normal and rainfall was reduced by one-half of the average in Europe since 1850. That summer's temperatures simulated temperature regimes that could be expected with long-term global warming.

A study was led by Philippe Ciais of the Laboratory of Environmental and Climate Sciences and French National Center for Scientific Research (Centre d'Etudes Atomiques-Centre National de la Recherche Scientifique) to quantify carbon storage and gas emission on grasslands. The study determined that throughout Europe the 2003 heat wave reduced the net storage of European ecosystems by 500 million tons (454 million metric tons) of carbon, or the equivalent of three years of greenhouse gas emissions by a country the size of France. The study illustrated that global warming could self-accelerate. By lowering the role of ecosystems as carbon sinks, the *greenhouse effect* would be accelerated and lead to increased warming. Ciais also points out that there are still many uncertainties that need to be answered.

Droughtlike conditions are not new to North America and the grasslands there. After the 1930s dust bowl era, where much of the Great Plains region suffered through extreme drought conditions, other areas have sporadically been affected as well, such as the southwestern United States over the past five years. In the summer of 2000, severe drought also affected southeastern states (Georgia and Alabama), and large areas of Texas went more than 67 consecutive days without any rainfall.

According to NASA, with the acceleration of global warming and rising temperatures, regions are likely to see more extreme dry events. Severe drought kills crops and dries topsoil, allowing it to be carried away by the wind, and evaporates lakes and other sources of freshwater, leaving everything parched. Current research is focusing on the ability to predict the onset of drought weeks, months, or even a year before it occurs so that adaptations to a warmer environment can be made.

Grasslands experiencing drought become very dry and prone to wildfire conditions. Whether fires are caused by lightning or by humans, they spread quickly, often engulfing entire hillsides. The months of July and August are especially dangerous. These two signs in Utah during July festivities remind residents not to light fireworks and to remember to conserve water. *(Nature's Images)*

Rangelands cover about 50 percent of the world's land area, and they are ecologically as important as the rain forests. They are also in as much danger of degradation and destruction. The world's rangelands may play a significant role in buffering the effects of global warming. These areas are used for the grazing of cattle and sheep and are seen as a carbon sink by some global warming scientists. An international team of researchers led by the University of California, Davis, has focused its attention on the central Asian rangelands in an effort to determine whether this is true. Montague Demment, who is the director of the University of California, Davis, Small Ruminant/ Global Livestock Collaborative Research Support Program, says,

The Bengal tiger is an endangered species. As global warming continues, habitat loss is occurring for these magnificent felines because of disturbance and rising sea levels, which are contaminating freshwater sources needed for survival. *(NOAA)*

"Properly managed rangelands could have a relevance for global climate stability comparable to conserving the tropical rain forests. We may very well find that conservation and restoration of the rangelands, in the United States and internationally, have a significant impact on the world's 'carbon budget' and are critical factors in slowing global warming."

The research team is looking at Asia's 647 million acres (262 million ha) of rangeland to determine how to restore them to healthy, productive, functioning grasslands. If this goal is met, they believe that they can store enough carbon to equal a 30 percent reduction in the carbon emissions caused by all the people in Russia. They also make dire warnings that if grassland restoration does not happen, if rangeland degradation through cultivation or development takes its place, substantial amounts of CO_2 could be released into the atmosphere. Rangelands offer vital functions for ecosystem health, including many that are listed below.

A common use of the U.S. grasslands is for the grazing of cattle, especially in the Southwest. As the climate warms and the environment becomes drier, droughtlike conditions become the norm and the land becomes taxed by overgrazing. Many rangelands will experience desertification—a degradation of the land into an unproductive wasteland. *(Keith Weller, Agricultural Research Service)*

Protects fragile soils

Stores large amounts of CO_2

Provides habitat for wild flora and fauna

Acts as watersheds for large river systems

Provides recreational opportunities

Provides reserves for wildlife

Maintains atmospheric quality

Purifies air and water

Retains nutrients for plants

Provides medicinal plants

Provides timber

Provides germplasm for new and wild relatives of existing crop and pasture plants

Provides food for livestock

Provides a source for biofuels

Provides fibers for cloth (cotton, wool, leather, hemp, flax, cashmere)

Provides industrial products (waxes, dyes, vegetable fats and oils)

Provides cooking oils (plant and animal fats and oils)

Although the health and destruction of grasslands and rangelands do not appear in the news as often as the destruction of the world's rain forests, they are in danger from the effects of global warming, as well as from human interference. Not only do they face degradation, but also desertification. The world's grasslands are also predicted to migrate northward, shifting the rich farmlands found in the central

Grasslands are also important in the raising of sheep. This woman from the Navajo tribe in New Mexico maintains a herd and uses their wool to weave beautiful blankets. If global warming continues and the rangeland becomes an unproductive desert, it will destroy the economic futures of many people who depend on the land for their survival. *(Ken Hammond, United States Department of Agriculture)*

portion of the United States northward into Canada. Crop production may become more risky and difficult. As farming regions shift, there is no guarantee that soils would be conducive to the large-scale growth of agriculture needed to feed the world. By 2050, the world's climate could be so unstable that world food production would be reduced by 50 percent.

ADAPTATION

As climate changes, it is important for land managers to adapt to the changes as they happen and manage the land in a responsible way. One of the biggest obstacles from global warming is the threat of invasive species encroaching on native grassland species. Invasion of species can happen via seed deliverance by livestock, off-road vehicles, road maintenance crews, and outdoor recreationalists. Invasive species need to be dealt with immediately before they become established, and land managers need to put management plans in place in order to be prepared for these changes.

Grassland restoration is necessary in many areas of the world, such as Australia, California, and the intermountain West in the United States. Restoration can include conducting controlled burns (burning fires intentionally to eliminate dry easily flammable organic material), letting natural fires burn, removing livestock, removal of invasive species, and reseeding of native species. The field of grassland restoration is fairly new, and a trial-and-error interactive learning approach will be necessary to establish the right balance in order to successfully restore and maintain the composition and structure of the world's grasslands faced with climate change.

Impacts on
Polar Ecosystems

The effects of global warming will be felt most strongly in the Earth's polar regions. Already temperatures have climbed by 4°F (2.4°C) in some polar regions, whereas the global average is 1°F (0.6°C) over the past century. Weather conditions are so harsh in polar ecosystems that a delicate balance must be maintained. Because of this sensitivity, polar regions are the first to show warning signs of global warming, such as an acceleration in the melting of glaciers.

This chapter will explore the ecosystems of both the Arctic and Antarctic regions and the impact that global warming is having on them. It will also address possible strategies of adaptation. Even though the polar areas are remote and do not support large centers of human settlement, understanding what is happening to these areas is important because what happens to the polar areas is important to the Earth's ecosystem and affects human settlements along coastal regions worldwide. Therefore, knowledge of the responses of polar regions to global warming can serve as an early warning system, giving people time to react and plan.

ARCTIC ECOSYSTEMS

In the Arctic, climate change is expected to be rapid and extensive. As temperatures warm up and ice melts, the Arctic will see serious changes over the next century. There is extensive sea ice at the periphery of the Arctic Ocean that forms and melts each year. These waters are important to the fishing industry, accounting for almost half of the total global production. Glacial decline, melting sea ice, rising sea levels, impacts on wildlife habitat, melting permafrost, impacts on Native tribal inhabitants, and changes to plant life are all being affected in the Arctic today because of global warming.

Water and Ice

Average temperatures in the Arctic region are rising twice as fast as they are in other places in the world. The ice is also getting thinner and thus more prone to breaking. The Ward Hunt Ice Shelf, located on the north coast of Ellesmere Island, Nunavut, Canada, is only one example—it was once the largest single block of ice in the Arctic. About 3,000 years old, it began cracking in the year 2000. By 2002, it had broken all the way through. Today it is breaking into pieces. The breakup of the Ward Hunt Ice Shelf has already caused major problems. It has forced polar bears, whales, and walrus to change their migration and feeding patterns, Native people are having to change their hunting territories, and coastal villages are being flooded.

There are about 2,000 valley glaciers in Alaska. Today less than 20 are still advancing, according to Bruce Molina, a geologist at the United States Geological Survey (USGS). Instead, nearly all of the glaciers in Alaska are melting. In the past seven years, the melting rate has been roughly twice what it had been previously. According to Molina, a striking example of glacial retreat that started as the Little Ice Age began to wind down and continues today is found at Glacier Bay, a popular destination for Alaskan cruise ships.

"Ironically, the climate event that made cruising into Glacier Bay not only possible, but popular, could ultimately take away its top attraction as many tidewater glaciers now retreat out of the water," Molina points out.

According to a report in the *Washington Post,* scientists at the National Oceanic and Atmospheric Administration (NOAA) say that the Arctic ice cap is melting faster than scientists had originally expected and will most likely shrink 40 percent by 2050 in most regions. This will devastate wildlife populations, such as polar bear, walruses, and other marine animals. They also predict that Arctic sea ice will retreat hundreds of miles farther from the coast of Alaska in the summer. While that could be good news in the short term for fisherman and open shipping routes and new areas for oil and gas exploration, it will spell disaster for the wildlife species that inhabit the region and ultimately for us all.

In 2001, the Intergovernmental Panel on Climate Change (IPCC) predicted that there will be "major ice loss by 2100." Then, in 2007, when they issued their next report, they remarked that "without drastic changes in greenhouse emissions, Arctic Sea Ice will 'almost entirely' disappear by the end of the century."

James Overland, an oceanographer at NOAA says, "The amount of emissions we have already put out in the last 20 years will stay around for 40 to 50 years. I'm afraid to say that a lot of impacts we will see in the next 30 to 40 years are pretty much already established." When asked about the effects on wildlife, he said, "It will have a profound effect on the animals that use sea ice all the time, including walrus and polar bears and ringed seals. You will actually have a change in the whole ecosystem. You will have winners and losers. Crabs, clams, walrus, and bears will not do well. Salmon, pollock, and others that live higher up in the water column will extend their range."

According to a report published May 1, 2007, in the *New York Times,* a study shows that climate scientists may have significantly underestimated the power of global warming from human-generated heat-trapping gases to shrink the sea ice in the Arctic Ocean. Dr. Julienne Stroeve, a researcher at the National Snow and Ice Data Center in Boulder, Colorado, discovered that since 1953 the area of sea ice in September has declined at an average rate of 7.8 percent per decade.

"There are huge changes going on," Dr. Stroeve says. "Just with warm waters entering the Arctic, combined with warming air temperatures, this is wreaking havoc on the sea ice."

Another study in LiveScience predicted that the Arctic summer could be completely ice free by 2105. According to the study, the Arctic has not been without ice for the past million years. Today, however, documented melting is accelerating. Jonathan Overpeck, of the National Science Foundation's Arctic System Science Committee, remarked: "What really makes the Arctic different from the rest of the non-polar world is the permanent ice in the ground, in the ocean, and on land. We see all of that ice melting already, and we envision that it will melt back much more dramatically in the future, as we move toward this more permanent ice-free state."

Flooding and melting ice have already begun to have an impact on the coastal areas in the Arctic where indigenous people live, such as in Alaska, Canada, Greenland, Russia, and Scandinavia. The melting in the Arctic has worldwide implications. When the ice melts, sunlight is no longer reflected back into space. This allows the Earth's surface to absorb more of the Sun's energy and heat up. This then starts a chain reaction of effects—the more it absorbs, the more it heats, and the more it melts; the more it absorbs, the more it heats, and the more it melts; and so on.

The warmer it gets, the more the glaciers melt, which adds freshwater to the ocean and upsets the Earth's energy balance, ocean circulation patterns, and ecosystems. Some estimates of global sea-level rise from Arctic melt have been placed at three feet by 2100. According to the U.S. Environmental Protection Agency (EPA), this increase would drown roughly 22,400 square miles (58,016 km²) of land along the Atlantic and Gulf Coasts of the United States—specifically in Texas, Florida, North Carolina, and Louisiana.

Warming in the Arctic will also affect weather patterns and food production worldwide. An example of this is given by NASA: In one of their computer models, they calculated that Kansas would be 4°F (2.4°C) warmer in the winter without Arctic ice, which usually sends cold air masses into the United States. Without these Arctic-induced cold air masses, winter wheat cannot be grown and soil would be 10 percent drier in the summer, wreaking havoc on summer wheat crops.

Arctic ponds that have existed for millennia are now disappearing due to global warming. This was discovered by John Smol from Queen's University in Ontario, Canada, and Marianne Douglas from the University of Alberta, Canada. They sampled 40 Arctic ponds in Canada every few years from 1983 to 2006 and found that the ecologically diverse water bodies were shrinking and getting saltier because hotter temperatures were causing the water to evaporate. Also alarming, they noted that wetlands originally surrounding some of these ponds had also been negatively affected and dried up, posing a wildfire threat under a lightning strike. Another negative impact of these drying wetland areas is that they stop being carbon sinks and become sources of carbon dioxide (CO_2) released back into the atmosphere.

In another study reported in LiveScience, 125 large Arctic lakes have completely disappeared in the past two decades because of rising temperatures. Lead researcher Laurence Smith of the University of California, Los Angeles, believes that global warming is thawing permafrost, and, when it melts, there is nothing to prevent the lake water from percolating down through the soil to the aquifers below. He says the changes happen abruptly: "From what we can tell from space [satellite imagery], a lake is either just fine or it's gone."

He reports that the sudden draining could alter entire continental ecosystems, affecting birds and other wildlife that depend on the waterways. For instance, migratory birds count on the lakes during the summer months as a food source for their young. "The loss of these lakes would be an ecological disaster," he said, referring to the thousands of ponds, lakes, and wetlands that exist in the Arctic during the summer.

ARCTIC IMPACTS

In 2005, the ice cap covering the Arctic Ocean melted to a size smaller than it had been since records began to be kept a century ago. In Greenland, glaciers are melting and traveling faster to the ocean, adding freshwater. These changes in the Arctic have the potential to affect the entire Earth. One of the most serious consequences could be the disruption of the thermohaline circulation. Called the *Great Ocean Conveyor Belt,* it brings warm, salty water from the *Gulf Stream* northward, where it releases heat to the atmosphere in the winter and warms the

North Atlantic region's climate. This is why Europe is warmer than it would normally be, considering its extremely northern location. Once the water reaches its northernmost trek in the Norwegian Sea, it sinks and then heads south to the equator in an endless loop. This current can be interrupted, however, if too much freshwater (such as from melting glaciers in the Arctic) is added to the ocean. Freshwater does not sink (because it contains no salt—dense salty ocean water sinks); so if freshwater joins the northern reaches of the current, it will prevent it from sinking, effectively shutting the current off and stopping the global transfer of heat, which could subsequently plunge Europe into an ice age.

The Arctic Council, an intergovernmental forum made up of eight Arctic nations (Canada, Denmark/Greenland/Faroe Islands, Finland, Iceland, Norway, Russia, Sweden, and the United States) and six indigenous people's organizations, had an assessment prepared by the Arctic Monitoring and Assessment Programme (AMAP), the Conservation of Arctic Flora and Fauna (CAFF), and the International Arctic Science Committee (IASC) of impacts on the Arctic as a result of global warming. They support the notion that the Arctic is extremely vulnerable to observed and projected climate impact and that it is currently facing some of the most rapid and severe climate change on Earth. Over the next 100 years, they report that climate is expected to change and these changes in the Arctic will be felt worldwide. While some of these changes are those caused by nature, the overwhelming trends and patterns are a result of human influence, resulting from the emission of greenhouse gases, such as carbon dioxide (CO_2).

CO_2 concentrations will remain at elevated levels for centuries, based solely on the CO_2 that has already been released. The lifetime of various greenhouse gases can range from decades to centuries. Because of this, some future warming is already inevitable. Future warming can be reduced, however, if emissions are limited now.

Although some areas show a warming trend, and some small areas even show a cooler trend, for the Arctic overall there is a clear warming trend already in progress. In many Arctic locations, winter temperatures are rising faster than summer temperatures. In fact, according to the Arctic Council's Arctic Climate Impact Assessment, in Alaska and western Canada, winter temperatures have increased as much as 5–6°F (3–4°C) in the last 50 years. Over the next 100 years, using a moderate emissions model, annual temperatures are projected to rise 5–8°F (3–5°C) over land and 11–17°F (7–10°C) over the oceans.

One of the biggest key climate change indicators in the Arctic is the declining sea ice. Researchers actually use the quantitative amounts of Arctic sea ice as an early warning system on global warming. According to the Arctic Research Center, over the last 30 years, the average sea ice extent has decreased by 80 percent, which is the equivalent of roughly 386,100 square miles (1 million km^2). As a comparison, this represents an area even larger than the size of Texas and Arizona combined. Of even more concern, the melting rate is accelerating. The most pronounced season of melting is in the late summer. It is projected that

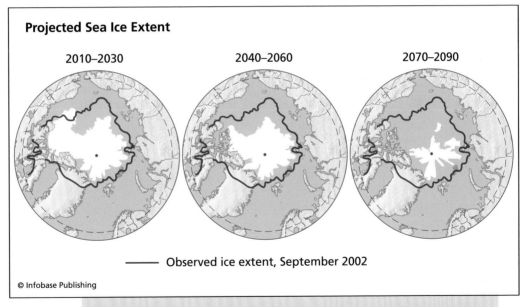

Projected Sea Ice Extent

2010–2030   2040–2060   2070–2090

——— Observed ice extent, September 2002

© Infobase Publishing

Projected sea ice extents for three time periods in the future. Sea ice is already declining today and is projected to decline even more rapidly in the future. These graphics of sea ice projections were derived from five climate models. As the century progresses, ice moves away from the coasts of Arctic landmasses. Some models project complete loss of summer sea ice in this century.

by 2100, late-summer sea ice will have decreased somewhere from 50 percent to complete disappearance.

Another major global impact that will occur as a result of ice sheet melting in the Arctic is sea-level rise. The Greenland Ice Sheet is the largest ice sheet in the Arctic. Its surface-melt area increased by 16 percent from 1979 to 2002, which is an area about the size of California and Massachusetts combined. In 2002, the total area of surface melt on the ice sheet broke all records kept to date—melting reached 6,600 feet (2,000 m) in elevation. Satellite data is used to monitor the surface of the ice sheet, and the sophisticated instruments on board can accurately measure slight changes in elevation, enabling scientists to accurately assess melt rates. Satellite data has shown an increasing melt rate since 1979. The only year the rate slowed down was in 1992. This was due to the eruption of Mount Pinatubo. The violent volcanic eruption ejected massive amounts of particles into the atmosphere, which reduced the amount of sunlight able to reach the Earth's surface, thereby cooling global temperatures that year.

Sea level rose roughly three inches (eight cm) over the past 20 years, and, as temperatures rise, the rate is increasing. This is due to two factors: melting ice and *thermal* expansion of seawater. The global average sea level is projected to rise anywhere from four inches to three feet (10–90 cm) by 2100. Some climate models have predicted that the Greenland Ice Sheet will completely melt, causing sea levels to rise 23 feet (seven m). As a comparison, a 1.5-foot (50 cm) rise in sea level will typically cause a shoreward retreat of coastline of 150 feet (50 m) if the land is relatively flat. This would flood areas, causing significant environmental and economic impacts.

Forest fires and insect infestations are also expected to increase as global warming intensifies. In terms of boreal forests burned in North

(opposite page) This shows the extent of the melted ice in Greenland in 1992 compared with 2002. In just 10 years, the amount of ice is one of the most rapidly retreating ice sheets being measured today. As the ice melts, it raises sea levels. The map of Florida illustrates the future for the coastal areas of Florida if sea level were to rise just 3.3 feet.

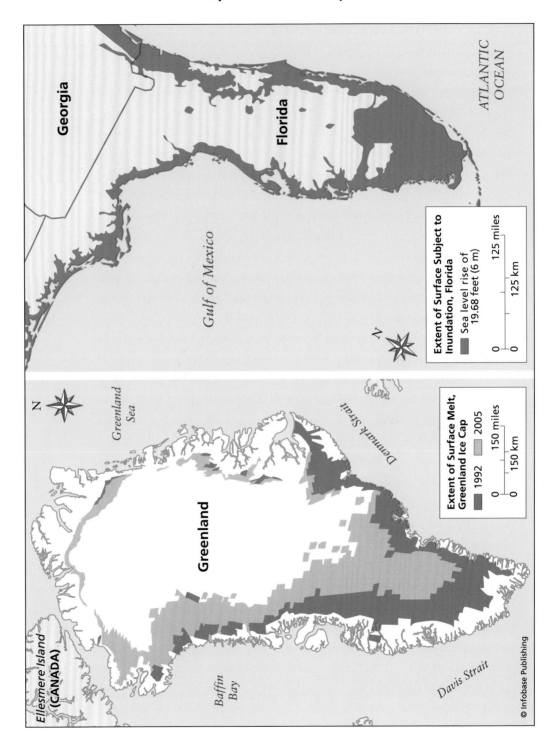

Extent of Surface Subject to Inundation, Florida

Sea level rise of 19.68 feet (6 m)

Georgia

Florida

Gulf of Mexico

ATLANTIC OCEAN

Extent of Surface Melt, Greenland Ice Cap

1992   2005

Ellesmere Island (CANADA)

Greenland

Greenland Sea

Denmark Strait

Baffin Bay

Davis Strait

© Infobase Publishing

America, the average area has more than doubled since 1970—which coincides with the documented warming of the region. Each year, fire fighters from the lower 48 states of the United States are flown to Alaska to fight these hot, fast, devastating wildfires.

Insects have also done great damage in Arctic ecosystems. In Abisko, Sweden, in 2004, moths destroyed extensive birch forests. In the Yukon Territory, from 1994 to 2002, spruce bark beetle infestations wreaked havoc on 741,316 acres (300,000 ha) in the Alsek River corridor in Kluane National Park in the Shakwak Valley north of Haines Junction. This infestation was the largest and most destructive outbreak of spruce bark beetles ever documented in Canadian forests. In the Kenai Peninsula in Alaska, during the 1990s, the world's largest outbreak of spruce bark beetles took place. Since 1989, more than 4 million acres (1.6 million ha) of white spruce and Sitka/Lutz spruce forest have had 10 to 20 percent of their trees die.

Human health has also been affected in the Arctic. Rural residents in small, isolated communities with few amenities and health support systems will be most at risk, as will people who depend on subsistence hunting and fishing for their food source. As climate forces animals to migrate or perish, indigenous people will lose a food source, leaving them with little to eat and limited contact with the outside world. Infectious diseases such as West Nile virus will also spread as new animals migrate into new areas. Already, West Nile virus has spread northward. By 2002, it was detected in 43 states in the United States and six Canadian provinces, spread by mosquitoes and migratory birds (West Nile virus originated in tropical Africa). The virus has adapted to many North American mosquitoes and more than 110 species of North American birds, some of which migrate to the Arctic. In the past, cold climate has limited the range of some insect-borne diseases, but now global warming and the ability for these robust diseases to adapt has allowed them to migrate, survive, and flourish even in Arctic climates.

Effects on Wildlife

Many species of wildlife are being negatively affected in the Arctic. Because these ecosystems are so fragile, habitat loss can occur rapidly, endangering the future of these animals and forcing them to face extinction. According to Greenpeace, many biologists doubt that Arctic animals will be able to adapt to climate change. Arctic ecosystems are already considered stressed, or inherently vulnerable, compared with temperate and tropical ecosystems. Their unique vulnerability is due to harsh Arctic weather. It is also due to the short and very sensitive nature of the Arctic food chain. Research suggests that this is where the Arctic differs from the temperate regions on Earth—animals here find it difficult enough to tolerate the stresses they are accustomed to, without trying to adapt to stresses of man-made origin. Arctic wild-life is more vulnerable to human-caused disturbance—such as climate change—than the populations of animals in other climatic regions. There are geographic limitations as well: The tundra is also the only major environment whose range is completely unable to shift north-ward in response to warming.

Robert Corell of the American Meteorological Society says, "The Arctic is experiencing some of the most rapid and severe climate change on Earth. The impacts of climate change on the region and the globe are projected to increase substantially. The Arctic is really warming now. These areas provide a bellwether of what's coming to planet Earth."

Several computer models of how the sea ice is shrinking have been examined by scientists working in Britain on the Arctic Climate Impact Assessment (ICIA). They have concluded that at the very minimum half the summer sea ice will disappear by 2100, with some models showing an almost total melt. The assessment states: "This is very likely to have devastating consequences for some Arctic animal species such as ice-living seals, walruses, and for local people for whom these animals are a primary food source. Should the Arctic Ocean become ice-free in sum-mer, it is likely that polar bears and some seal species would be driven toward extinction."

Polar Bears

Polar bears depend on sea ice for survival. They use the ice to hunt seals, and they use ice corridors for migration. If summer ice melts, the polar bears have nowhere to go and cannot survive. Global warming has already caused sea ice to thin to dangerous levels, and melting peri-ods are occurring two weeks earlier, robbing mother polar bears of the time they need to feed and build up the fat that enables them to sustain themselves and feed their young.

Warning signs of the effects of global warming on mother polar bears are already manifesting themselves. Polar bears are thinner overall than in the past, and the birth rate and survival of the cubs has decreased. The U.S. Fish and Wildlife Service (FWS) announced on February 8, 2006, that they will begin the process of evaluation in order to determine if the polar bear needs to be added to the official list of endangered species. This is "a significant acknowledgment of what global warming is doing to the Arctic ice," said Kassie Siegel, an attorney with the Center for Biological Diversity in Joshua Tree, California. In December 2005, this conservation group, along with Greenpeace and the Natural Resources Defense Council (NRDC), sued the U.S. government to protect the world's polar bears from extinction. A proposed rule that would add the polar bear to the federal list of threatened and endangered species was published on January 9, 2007, which opened a 90-day comment period on the proposed listing. On May 14, 2008, former secretary of the interior Dirk Kempthorne announced that he was listing the polar bear as a threatened species under the Endangered Species Act (ESA) on the recommendation of the FWS.

Polar bears are facing a tough situation right now. As the Arctic ice retreats, polar bears are struggling to survive. Lack of ice takes away valuable hunting grounds and migration corridors. *(Publitek, Inc.)*

THE BEAR FACTS

Here are some interesting facts about polar bears:

- Polar bears roam from Russia to Alaska, from Canada to Greenland, and onto Norway's Svalbard archipelago.
- Worldwide population is estimated at 20,000 to 25,000, of which about 60 percent is in Canada.
- Global warming is the main threat to polar bears today.
- In the 1960s and 1970s, hunting was the major threat to polar bears.
- Polar bears are also affected by pollution, drilling and mining, and fishing and hunting.
- Polar bears depend on a frozen platform from which to hunt seals. Without ice, the bears are unable to reach their prey.
- Compared to data collected 20 years ago, today's polar bears have lower cub survival rates. The weight and skull size of the adult males are also smaller today.
- In a warming Arctic, polar bears may not be able to swim the distances required to reach solid ice if reduced ice coverage causes bigger stretches of open water. In 2004, four polar bears drowned off the coast of Alaska while trying to swim to pack ice. The ice had retreated 160 miles (257 km). The polar bears were 60 miles (97 km) offshore—too far for them to swim. Rough seas and high winds made it impossible to reach their destination.
- Because polar bears are the top predator of the Arctic, changes in their distribution and population could adversely affect the entire Arctic ecosystem.

The time to begin planning for, and saving, the polar bear population is now.

In a study conducted by the Canadian Wildlife Service, they found that the bears' main food source, ringed seals, which live on the ice of Hudson Bay, are becoming less accessible because of a shorter ice season. They have determined that sea ice is melting there three weeks earlier, leaving the polar bears with less time to hunt, and that they are returning to the land in poorer condition. According to Ian Stirling, Nicholas J. Lunn, and John Iacozza, the scientists that conducted the study, the polar bears are starving due to global warming. According to them: "The Hudson Bay polar bears are unique in the Arctic because of their tendency to fast for six to eight months each year, depending heavily on hunting during the sea ice season for survival. Since the sea ice season is the shortest in Hudson Bay of all the regions in the Arctic Ocean, these bears are on the edge of survival, and are likely to be among the first to be affected by sea ice decline."

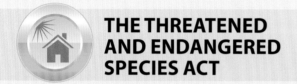

THE THREATENED AND ENDANGERED SPECIES ACT

Half of the recorded extinction of mammals over the past 2,000 years has occurred in the most recent 50-year period. The Endangered Species Act is a law that was passed by Congress and signed by President Nixon in 1973. It is regarded as one of the most comprehensive wildlife conservation laws in the world. In 1973, 109 species were listed. Today, more than 1,000 species are listed.

Congress had enacted two similar laws—one in 1966 and another in 1969—but neither did more than create lists of vanishing wildlife species. This was no better than publishing a list of murder victims but doing nothing to catch the murderers. The Endangered Species Act changed all that. It forbids people from trapping, harming, harassing, poisoning, wounding, capturing, hunting, collecting, importing, exporting, or in any other way harming any species of animal or plant whose continued existence is threatened. The FWS manages the listing of land and freshwater species. The Natural Marine Fisheries Service is in charge of ocean species.

This act protects all species classified as threatened or endangered, whether they have commercial value (people want to buy, sell, and collect them) or not. It also forbids federal participation in projects that jeopardize listed species and calls for the protection of all habitat critical to listed species. Assistance is provided to states and foreign governments to ensure this protection.

Under this law, "endangered" designates a species in danger of extinction throughout all or a significant part of its range. "Threatened" refers to

Kevin Jardine, of Greenpeace, adds "We're wrong if we think that climate change is something that will happen far off in the future. Polar bears are starving now and we need to act now to stop climate change. Starving polar bears are a major climate danger signal."

In an article in *USA Today* on June 13, 2006, entitled "Study: Polar Bears May Turn to Cannibalism," it was reported that polar bears in the southern Beaufort Sea may be turning to cannibalism because longer seasons without ice keeps them from getting to their natural food. In April 2004, there was a documented report of the first-ever killing of a

species likely to become endangered in the foreseeable future. The FWS also maintains a list of "candidate" species. These are species for which FWS has enough information to warrant proposing them for listing as endangered or threatened, but they have not yet been proposed for listing.

The law's ultimate goal is to recover species so they no longer need protection under the Endangered Species Act. The law provides for recovery plans to be developed describing the steps needed to restore a species to health. Public and private agencies, institutions, and scientists assist in the development and implementation of recovery plans.

The Endangered Species Act is the law that puts into effect U.S. participation in the Convention on International Trade in Endangered Species of Wild Fauna and Flora (CITES)—a 130-nation agreement designed to prevent species from becoming endangered or extinct because of international trade.

There are some exceptions to the protection provided by the Endangered Species Act. Natives of Alaska can hunt endangered animals for use as food or shelter. An endangered animal can also be killed if it threatens the life of a human or if it is too sick to survive.

Conservationists agree that it is in the best interest of mankind to minimize the losses of genetic variations. All life on Earth is an important resource. Many species—even those unknown at this point—could provide answers to questions and solutions to problems in the future. If humans destroy species now, they may never understand the potential of these species as natural resources.

female in a den shortly after it gave birth. In the study, global warming was blamed for the behavior—the polar bears are starving.

In February 2005, the Center for Biological Diversity (one of the groups who petitioned the federal government to list polar bears as threatened under the federal ESA) also commented on the incidence of cannibalism. Kassie Siegel, lead author of the petition, said "Cannibalism demonstrates the effect on bears. It's very important new information. It shows in a really graphic way how severe the problem of global warming is for polar bears."

Deborah Williams of Alaska Conservation Solutions, a group that promotes solutions for global warming, remarked, "This study represents the 'bloody fingerprints' of global warming. This is not a Coca-Cola commercial. This represents the brutal downside of global warming."

Walrus

According to Discovery News, in October 2007, thousands of walrus showed up on Alaska's northwest coast in what conservationists are calling a "dramatic consequence of global warming melting the Arctic sea ice."

Generally, walrus are found on the Arctic ice pack. But because the ice cap has melted so much, it had retreated far to the north of the outer continental shelf where the walrus' food is normally found. Because their traditional source of food was out of reach, the walrus came ashore on Alaska's rocky beaches in search of food instead.

Tim Ragen of the Marine Mammal Commission says in response, "It looks like the animals are shifting their distributions to find prey. The big question is whether they will be able to find sufficient prey in areas where they are looking."

Several thousand walrus have abandoned the ice pack for areas along a 300-mile (483-km) stretch of shoreline along Alaska between Barrow and Cape Lisburne.

In the spring and summer of 2004, nine sightings of baby walrus swimming far away from shore were spotted, abandoned by their nursing mothers. Carin Ashjian, a biologist at the Woods Hole Oceanographic Institution, says, "The young can't forage for themselves and are dependent on their mother's milk for up to two years."

The young calves were too far from land and would eventually starve or drown. Researchers believe that the mothers had to swim farther and farther away from shore to find ice for the calves to rest on and eventually had to abandon them in waters too deep for the mothers to reach food. The researchers said that ice was "virtually absent" throughout the area where they saw the abandoned calves. The area was 53–134 miles (85–215 km) from shore in waters that were 9,842 feet (3,000 m) deep. None of the abandoned calves were recovered.

Another researcher, Lee W. Cooper, a biogeochemist at the University of Tennessee, remarked, "If walruses and other ice-associated marine mammals cannot adapt to caring for their young in shallow waters without sea ice available as a resting platform between dives to the seafloor, a significant population decline of this species could occur."

Seals

Seals—ringed, ribbon, and bearded—are also ice-dependent animals and very vulnerable to reductions in sea ice. They give birth to their

Seals are also having problems in their polar habitats due to lack of ice cover. As global warming continues, it will be harsh for them to survive in their environment. This is a baby Weddell seal in Antarctica. *(Commander John Bortniak, NOAA Corps Collection)*

TESS

Before a plant or animal species can receive protection under the Endangered Species Act, it must first be placed on the federal list of endangered and threatened wildlife and plants, maintained by the FWS. The listing program follows a strict legal process to determine whether to list a species, depending on the degree of threat it faces. An 'endangered' species is one that is in danger of extinction throughout all or a significant portion of its range. A 'threatened' species is one that is likely to become endangered in the foreseeable future. FWS also maintains a list of plants and animals native to the United States that are considered as candidates or proposed for possible addition to the federal list. All of the FWS's actions—from proposals, to listings, to removals (delisting)—are announced through a publication called the *Federal Register*.

TESS—Threatened & Endangered Species System—is the database where all these categories are listed.

Candidate species are those plants and animals that FWS has enough information regarding their biological status and threats in order to propose them as endangered or threatened under the Endangered Species Act, but are not listed yet because other species are seen as a higher priority. A species is added to the list when it is determined to be endangered or threatened because of any of the following factors:

- The present or threatened destruction of its habitat or range
- Its overuse for commercial, recreational, scientific, or educational purposes
- Disease or predation
- If existing regulations are not protecting it
- Other natural or manmade factors that affect its survival

pups on the ice, nurse them on the ice, and use the ice as a resting platform. They also forage for food near the ice's edge and under it. Without the availability of ice, seals cannot exist in the Arctic habitat. Thus, as global warming continues to intensify, their existence is threatened.

In order to list a species, several steps must be followed. The first step is to publish a "notice of review" that assesses the condition of a species. If the species meets the criteria, it becomes a candidate for listing. Through notices of review, the government collects information that will allow the species to be evaluated. The reviews are posted in the *Federal Register.*

Because of the large number of candidates and the time required to list a species, FWS has developed a priority system designed to prioritize the plants and animals according to the greatest need for preservation.

Once a species is proposed in the *Federal Register,* any interested person can comment and provide additional information on the proposal—usually during a 60-day review period—and submit statements at any public hearings that may be held. In order to inform the public of these special hearings, news releases and mailings are issued.

Once the government receives public comments, they are analyzed and considered in the final rule-making process. Within one year of a listed proposal, one of three possible courses of action is taken:

- A final listing rule is published.
- The proposal is withdrawn because the biological information does not support the listing.
- The proposal is extended for up to six months due to substantial disagreement within the scientific community concerning the need to list a species. After the six months, a determination is made.

When a species is approved, the final listing rule becomes effective 30 days after publication in the *Federal Register.* After a species is listed, its status is reviewed at least every five years to determine if federal protection is still necessary.

PERMAFROST

Roughly 24 percent of the land area in the Northern Hemisphere covers perennially frozen ground called permafrost. Another 57 percent freezes seasonally. With the increase of global warming, this permanently frozen ground is starting to thaw and the seasonally frozen ground has decreased by 15 to 20 percent during the 20th century.

According to Tingjun Zhang of the University of Colorado at Boulder, the warm-up in locations all across Russia has been documented as a one-degree increase in the average temperature of soil 16 inches (40 cm) below the surface. Zhang says, "The change is real. It is happening."

The thaw is not just happening far up north. It is also occurring in areas of the United States. In fact, roughly 80 percent of the soil in the United States freezes each winter. Changes to this cycle because of global warming will affect crops, native plants, and even how much carbon is exchanged with the atmosphere. Frederick Nelson, a geographer at the University of Delaware, says, "There is widespread evidence that global warming is responsible for the observed changes in seasonally frozen soil and permafrost."

Because water in the soil expands when it is frozen and shrinks when it thaws, it causes uneven movements in the ground. As global warming persists, thawing of permafrost will cause the ground to shift enough so that it will be a disaster to structures built on the ground's surface, such as buildings, roads, pipelines, airports, industrial facilities, and railroad tracks. Thawing of permafrost has also contributed to landscape erosion, slope instability, and landslides. Coastal erosion up to 100 feet (30 m) per year has been observed in areas of the Siberian, Canadian, and Alaskan Arctic. A small Inuit village of 600 homes in the Chukchi Sea area is currently at risk of falling into the sea. As bad as all this is, there is another huge problem with melting permafrost—it releases methane, a potent greenhouse gas, that has been trapped in the permafrost for thousands of years.

SHIFTING VEGETATION ZONES

Global warming is also causing vegetation zones to shift as vegetation types attempt to migrate northward toward cooler climates. Forests are expected to migrate into the Arctic tundra. The tundra will shift into the polar deserts. These changes are expected to eliminate the tundra ecosystem to the lowest extent it has been in the past 21,000 years. This will have a great impact on the breeding area for many birds, as well as grazing areas for land animals.

It is also expected that the reduction in tundra and the related expansion in forest will cause a decrease in surface reflectivity because forests are darker and have more textured surfaces, causing them to absorb more solar radiation than tundra, which is lighter, smoother, and more reflective. Expanding forests will also mask highly reflective snow. This will then cause a positive *feedback,* whereby warming will lead to more forest growth, which will cause more warming.

There is positive news to report in all this negativity. Some Arctic species such as white arctic bell-heather have been able to adjust to the changes in climate by recolonizing the islands of the Svalbard archipelago in the Arctic Ocean about midway between mainland Norway and the North Pole. These islands, in the past 20,000 years, have undergone warm and cold spells, and species have been able to adapt. This gives hope that other flora and fauna will be able to adapt and shift also. According to Julie Brigham-Grette at the University of Massachusetts, "As the proper habitat is available, plants will survive. If dispersal is not a limiting factor, then maybe the rate of warming ongoing in the Arctic will not be a limiting factor in plant survival in distant places." Some view this ability to disperse widely as an evolutionary consequence of the region's ability to adjust to sharp climatic shifts.

ANTARCTIC ECOSYSTEMS

The Antarctic Peninsula is warming faster than anywhere else on Earth, and its glaciers are in active retreat. The IPCC has projected sea levels to rise 7–23 inches (18–59 cm) in this century from temperature rises of 2.9–5.7°F (1.8–4°C). The fringe areas—islands—around the polar region are also being affected. The Heard Island glacier, 620 miles (1,000 km) north of Antarctica, is also melting.

As pieces of the ice cap break off into the ocean, it creates the same effect as uncorking a bottle. By removing the ice—which blocked the flow of glaciers outward into the sea—it now allows the glaciers to freely flow out into the sea at an accelerated rate.

The melting of glaciers into the sea contributes to sea-level rise. Steve Rintoul of Australia's Commonwealth Scientific and Industrial Research Organisation (CSIRO) Marine Research says, "About 100 million people around the world live within 3 feet (1 m) of the present-day sea level. Those 100 million people will need to go somewhere." As a comparison, every yard of sea-level rise causes an inland recession of around 300 feet (100 m). In addition, more erosion occurs with every storm.

Scientists at the National Snow and Ice Data Center (NSIDC) report that after the Larsen B Ice Shelf collapsed in 2002, interior glaciers began to flow outward to the ocean faster. Scientists realized that the ice shelves in Antarctica partially served to hold back the continent's interior glaciers.

The effects of this intensified cycle of melting will be felt worldwide. With rising seas, coastal areas will be highly vulnerable to flooding. Freshwater wells and other resources may be degraded or destroyed. Entire low-lying islands may completely disappear.

WILDLIFE

According to National Geographic News in August 2001, scientists believe that global warming caused the unexpected collapse in the numbers of minke whales found in the Antarctic. They believe that because of ice loss half the whale population has been lost during the past decade. The situation has certainly gotten the attention of con-servationists and those demanding that nations implement the Kyoto Protocol—the international treaty devised to fight climate change, implementing bans and restrictions, as well as solutions.

Although it is very difficult to count whales at sea, when minke whale populations were calculated between 1985 and 1991, best esti-mates put them at 760,000. During later counts in the 1990s, data sug-gested that only 380,000 remain. Global warming is being blamed. One explanation for their dwindling numbers is that the ice has retreated so far the whales are having a difficult time feeding on the krill they need in order to survive. To support this theory, the Australian govern-ment has determined that the ice has dropped by one-fourth between the mid-1950s and early 1970s. Sidney Holt, a member of the Scien-tific Committee between 1960 and 1997, says, "Global warming is the 'likeliest hypothesis' for the crash." The Scientific Committee is an orga-nization that serves under the New South Wales (NSW) government Department of Environment and Climate Change. It was established under NSW's Threatened Species Conservation Act and is an independent organization of scientists who are directly appointed by the minister for climate change and the environment.

The members are experts in their respective fields and cover diverse scientific areas such as vertebrate biology, plant biology, invertebrate biology, limnology (the study of freshwater bodies), plant community ecology, terrestrial ecology, aquatic biology, and genetics.

The committee serves several critical functions, including deciding which species, populations, and ecological communities should be listed as critically endangered, endangered, or vulnerable. They also decide which threats to native plants and animals should be declared key threatening processes.

One of their branches is the Scientific Committee on Problems of the Environment (SCOPE). This is an interdisciplinary body of natural and social science expertise that is focused on global environmental issues. It focuses on the interactions between the scientific community and the decision- and policy-making components of society in an effort to bridge the gaps between the two and promote practical solutions to problems that not only help society but protect the environment. It deals not only with current environmental issues but also future ones. Topics include global warming and climate change, pollution, environmental hazards, the effects of human activity on the environment, land use decisions, biodiversity, and ocean and watershed ecology.

There is also a Scientific Committee on Oceanic Research (SCOR). SCOR activities focus on the promotion of international cooperation in planning and conducting oceanographic research, as well as solving problems that may stand in the way of accomplishing necessary research. SCOR is made up of approximately 250 scientists from 35 nations. Because SCOR is not a government-funded organization, it receives funding from a variety of private sources and relies heavily on international cooperation and support in order to function, conduct oceanographic research, and solve methodological and conceptual problems.

SCOR has been involved in several large-scale ocean research projects, beginning with the International Indian Ocean Expedition in the early 1960s. Since then, it has been involved in the Global Ocean

Ecosystem Dynamics project, the Joint Global Ocean Flux Study, the Tropical Ocean-Global Atmosphere project, the World Ocean Circulation Experiment, and the Global Ecology and Oceanography of Harmful Algal Blooms program. SCOR and the Intergovernmental Oceanographic Commission (IOC) have cosponsored activities related to the ocean's role in global climate change since 1979.

There is also the Scientific Committee on Antarctic Research (SCAR). Its function is to initiate, develop, and coordinate international scientific research in the Antarctic region and provide critical information as to the role Antarctica plays in the overall Earth ecosystem. It also provides objective and independent scientific advice to the Antarctic Treaty Consultative Meetings and other organizations on issues of science and conservation affecting the management and well-being of Antarctica and the Southern Ocean.

Penguin decline in Antarctica has also been linked to climate change. According to a report in National Geographic News, over the past 50 years, the population of Antarctic emperor penguins has declined by 50 percent. During a long warming spell in the Southern

The penguins at the South Pole may have their habitat destroyed as the ecological systems in the polar regions begin to change under hotter temperature regimes. Found only in Antarctica, these emperor penguins live in the southwest Ross Sea area. *(Michael Van Woert, NOAA Corps Collection)*

Ocean during the 1970s, there was a notable decline in the emperor penguin population at Terre Adélie, Antarctica.

Similar to the Arctic, this decline relates to the lack of sea ice. As the atmosphere and sea-surface temperatures rise, the amount of ice in the ocean declines. As a result, there are less krill to populate the area. Krill are a staple of the emperor penguin's diet, and, thus, with a greatly diminished food source, the penguins' mortality rate increases. According to Henri Weimerskirch of the Centre national de la recherche scientifique, "The population decreased because of the low rates of survival more than four to five successive years."

One of the life-forms in the Antarctic that may be able to give scientists valuable clues about global warming are microscopic worms called nematodes. These worms exist beneath the rocky, frozen polar soil, where they help boost soil fertility by accelerating decomposition and decay. They recycle nutrients and make them available to nourish plants. Because these microscopic organisms form the basis of a food chain, it is important to understand how they will react to warmer global temperatures. When conditions get cool enough, the worms become dehydrated and go dormant. Once summer comes and the environment warms up, the worms wiggle back to life.

Carbon exchange and storage are other concerns. In the McMurdo Dry Valleys in Antarctica, where research is being conducted, the nematodes play a disproportionate role in recycling carbon from one useful form to another. Although they only comprise 0.025 percent of the total carbon themselves, they are responsible for approximately 10 percent of the total carbon processed in the area. The thing that has Dr. Diana Wall, soil ecologist from the Natural Resource Ecology Laboratory at Colorado State University, and her colleagues concerned is that in the past 10 years, while temperatures across Antarctica have been climbing, those in the Dry Valleys have experienced a cooling period. This cooling trend has led to cooler and drier soils, causing the nematode population to shrink 65 percent since 1993. The drop in population has resulted in the loss of a third of the total carbon cycling in the ecosystem.

According to Dr. Wall, "What we're seeing is that the beast that has the most to do, or a disproportionate amount to do, is one species that is crashing and burning."

Dr. Ross Virginia, another researcher on the team, remarked, "The complexity in most other soils has prevented one from teasing apart these variables in a way that we've done down there. But the effects on biodiversity in the cold-desert ecosystem of the Dry Valleys make an apt comparison to regions like the hot, arid desert of the Sahara, which is increasingly under threat from desertification and climate change. And such areas make up about one-third of the Earth's land surface."

On a global scale, the amount of CO_2 released by soil invertebrates and microbes is greater than 10 times the annual carbon emissions from fossil fuels. The oceans and forests perform crucial roles as carbon sinks, drawing in carbon that would otherwise reside in the atmosphere. Soils hold more carbon than trees and the atmosphere combined. The nematodes, therefore, could play a significant role in the global warming scenario by affecting carbon exchange and storage. Future cooling or warming trends may affect the outcome of nematode mortality and CO_2 levels.

Dr. Wall is currently conducting studies of these worms to determine how temperatures and other factors affect them. Since 1993, she has operated 32 worm farm chambers. Each chamber creates a greenhouse effect, raising the temperature of the soil about 3.6°F (2°C). To date, the warmer temperatures in the chambers have also led to a decrease in nematodes. Therefore, it is possible that global warming temperatures could threaten nematodes and other microscopic organisms in the soil web—permanently changing the chemistry and nutrient content of the soil. The effect this would also have on the food chain is still unknown, but it is one that has biologists, ecologists, and climate scientists concerned.

ADAPTATION

According to the IPCC's Third Assessment Report in 2001: "Changes in climate that have already taken place are manifested in the decrease in extent and thickness of Arctic sea ice, permafrost thawing, coastal erosion, changes in ice sheets and ice shelves, and altered distribution and abundance of species."

In the IPCC's Fourth Assessment Report in 2007, they stated:

- There is high confidence that recent regional changes in temperature have had discernible impacts on many physical and biological systems.
- With regard to changes in snow, ice, and frozen ground (including permafrost), there is high confidence that natural systems are affected. There is an enlargement and increased numbers of glacial lakes.
- There is increasing ground instability on permafrost regions and rock avalanches in mountain regions.
- There are changes in some Arctic and Antarctic ecosystems, including those in sea-ice biomes, and also predators high in the food chain.
- There is increased runoff and earlier spring peak discharge in many glacier and snow-fed rivers.
- The warming of lakes and rivers in many regions will alter the effects of thermal structure and water quality.
- There is high confidence, based on substantial new evidence, that observed changes in marine and freshwater biological systems are associated with rising water temperatures, as well as related changes in ice cover.
- Settlements in mountain regions are at enhanced risk of glacier lake outburst floods caused by melting glaciers.
- In the polar regions, the main projected biophysical effects are reductions in thickness and extent of glaciers and ice sheets and changes in natural ecosystems with detrimental effects on many organisms including migratory birds, mammals, and higher predators. In the Arctic, additional impacts include reductions in the extent of sea ice and permafrost, increased coastal erosion, and an increase in the depth of permafrost seasonal thawing.
- For human communities in the Arctic, impacts, particularly those resulting from changing snow and ice conditions, are projected to be mixed. Detrimental impacts would include those on infrastructure and traditional indigenous ways of life.
- Beneficial impacts would include reduced heating costs and more navigable northern sea routes.

- In both polar regions, specific ecosystems and habitats are projected to be vulnerable, as climate barriers to species invasions are lowered.
- Arctic human communities are already adapting to climate change, but both external and internal stressors challenge their adaptive capacities. Despite the resilience shown historically by Arctic indigenous communities, some traditional ways of life are being threatened and substantial investments are needed to adapt or relocate physical structures and communities.

The effects of global warming over the polar land areas have been greater than anywhere else in the world. As the snow and ice melt, the darker surfaces cause more of the Sun's radiation to absorb at the Earth's surface, heating it further. The warming that is occurring today in the polar regions agrees with the results of many of the climatic model simulations of changes due to human impacts and the concentration of greenhouse gases of the past century. Like a vicious circle, as warming accelerates in the polar regions, it triggers impacts over the rest of the Earth, such as rising sea levels due to the melting of glaciers and ice sheets. Some models have predicted the polar areas will warm by 5–10°F (3–6°C) by 2100.

Because much of the damage that has already been done through the emissions of CO_2 into the atmosphere over the past 200 years or so will continue to cause the Earth's atmosphere to warm for decades or more to come, humans must not only work on solving the problem, but must find ways to help ecosystems adapt to the changes that will happen as the Earth continues to warm. Helping fragile polar ecosystems adapt requires effective land management. According to climate specialists at the World Wildlife Fund (WWF), the most effective management strategy must be based on an ecosystem's natural resilience while reducing vulnerabilities. Resilience is the ability of an ecosystem to maintain itself and its functionality even if disturbance is occurring. Vulnerability is a measure of how susceptible an ecosystem is to the adverse impacts of change. According to the IPCC, some Arctic species, such as those that depend on sea ice (like the polar bear and walrus), are more vulnerable to the effects of climate change than others.

As far as adaptation, the polar environment is one of the most difficult to adapt because it is so fragile and the changes are happening so fast. Even in the face of this difficulty, however, the WWF has suggested the following measures to help preserve the polar habitats:

1. **Habitat protection:** There is a need to establish protected areas that prohibit human activities, such as industrial development or commercial activity. Terrestrial, coastal, and marine ecosystems must all be taken into account in terms of protection.

2. **Protection of species:** Humans must take into account not only where they exist now but, where they will migrate to as the climate warms.

3. **Appropriate management scale:** Habitats must be managed in their entirety so that they do not become fragmented. If a habitat becomes broken by towns or other human infringement and cannot migrate poleward, the habitat will be threatened.

4. **Habitat must be managed at all scales:** This includes local, regional, and circumpolar scales. This is necessary to help protect migratory species.

5. **Reduce other stresses that are not climate-related:** This would include reducing or eliminating activities such as overfishing and overhunting.

6. **Monitor shipping traffic:** As sea ice melts and sea routes open to shipping, it is important to monitor usage, limit pollution, keep ships away from highly sensitive ecological areas, and monitor traffic to avoid collisions or spills.

7. **Keep impacts from tourism to a minimum:** It is critical to monitor tourism in order to prevent damage to soils and vegetation, wildlife disturbance, and pollution.

The effects of global warming are already so pronounced in the polar regions that international action needs to be taken now. Without significant international cooperation, the polar areas will be hit hard and be transformed into unfamiliar places for many plant and animal species that will begin a journey toward extinction. If the polar regions are indeed an early warning measuring stick whose effects will touch other ecosystems, the time to respond in an environmentally responsible way is now.

Impacts to Desert Ecosystems

The world's principal deserts are found in two major zones: at 25–35° latitude North and South of the equator. Desert ecosystems are defined as regions that are very arid and dry, receiving less than 10 inches (25 cm) of rain a year. This means deserts are not confined to just hot, dry areas—they can also exist in the cold, dry areas within polar regions.

Because these regions are so physically harsh, the plants and animals that live within them have learned to adapt and survive in an extremely hostile environment. Covering about one-fourth of the Earth's land surface (20.9 million square miles [54 million km²]), they are dominated by bare soil and scarce vegetative cover.

By definition, deserts do not receive much rainfall. When rainfall does occur, it can fall in short periods of great, but brief, abundance. The plants and animals that survive in the desert have developed very specific adaptations allowing them to successfully live there. For example, they have learned to live in extreme temperatures and without much water. This is accomplished by physiological adaptations and by using the environment to their advantage (such as being active only during the cooler evening hours in the hot deserts).

As global warming continues, current deserts will be affected as temperatures rise, water becomes scarcer, and food sources become problematic. As it gets warmer and desertification spreads, ecosystems worldwide will feel a change. Understanding the fragile balances in desert ecosystems and how global warming will affect them is important as land managers look at future options. This chapter will look at the different types of desert habitats and the expected effects of global warming and then discuss the relevant issues of drought, loss of productivity, the desertification dilemma, deadly heat waves, and long-term concerns about wildfire. As current arable, productive lands face conversion to desertlike landscapes in the near future, it is important to understand the ramifications of the processes now in order to prevent extensive destruction.

DESERT HABITATS

For a region to qualify as a desert, the annual potential evaporation must be greater than the annual precipitation. The deserts of the world are often referred to as the drylands of the world. Roughly 300 million people live in the Earth's drylands today. There are four types of desert habitat: (1) hot and dry deserts, (2) cool coastal deserts, (3) semiarid deserts, and (4) polar regions.

The hot and dry deserts are those areas that experience rapid evaporation of any moisture received. The soils are very shallow and coarse, usually composed of gravel or rocks. Finer particles do not exist because they are easily blown away, leaving only the heavier particles behind. Because of their coarse texture, these soils are capable of holding and storing water reserves.

The plants that live in these habitats have adapted to the droughtlike conditions. Desert succulents, such as cacti, survive by accumulating moisture in their fleshy tissues. Their roots are shallow and concentrated near the surface so that when it does rain the plant can store the water.

Desert ecosystems are the driest ecosystems on Earth. Both plants and animals must learn how to adapt to this harsh environment. *(Nature's Images)*

Ephemeral species are also found in the desert. These species have a short life cycle, but leave behind hardy seeds to ensure their survival under extreme conditions. When growing conditions are right, desert ephemerals grow very fast and reproduce at a very high rate.

Cool coastal deserts exist at latitudes of 30° North or South, and their climate is influenced by a cool offshore ocean current. Cool winters are followed by moderately long, warm summers. The soils in these deserts contain more salts than those of other deserts. Semiarid deserts are those with a wide range of annual temperatures. In the peak of the summer, they are often 100°F (38°C), but in winter they can be as cold as 54°F (12°C). In the world's polar deserts, most of the moisture is locked away as ice.

Animals in desert habitats have had to adapt to the harsh conditions they live in. The more mobile species, such as birds and large mammals, can migrate along desert plains or into nearby mountains during times of extreme drought. Smaller animals typically seek cooler, shady places. There are many rodents, invertebrates, and snakes in desert habitats. They avoid daytime heat by staying inside the cooler caves, if present, or burrows, and hunting during the nighttime hours. Animals that must be active during the day regulate their activity and rest as much as possible in shady areas.

DROUGHT

Global warming is expected to cause intensification in the hydrologic cycle, with a marked increase in evaporation over land and water. The higher the evaporation rate, the greater the drying will be of soils and vegetation. According to the Union of Concerned Scientists (UCS), current climate models also project that there will be changes in the distribution and amount of rainfall. In areas where there is a decrease in rainfall and an increase in evaporation, droughts will occur, be more severe, and last longer.

Drought already threatens the lives of millions of people on Earth. As the atmosphere continues to heat from global warming, the negative effects and hardships brought on by drought will spread across the globe as areas heat up and dry out. According to a study conducted by the Hadley Centre for Climate Prediction and Research, as reported by the *Independent* in Britain in 2006, "Extreme drought, in which agriculture is in effect impossible, will affect about a third of the planet."

The Hadley Centre also added that their prediction may be an underestimate. It is widely recognized that the worst hit will be the large populations of the developing countries—those who can least afford to deal with the unfortunate consequences.

Andrew Pendleton of Christian Aid remarked at the Climate Clinic at the Conservative Party conference, where and when the findings were released, "This is genuinely terrifying. It is a death sentence for many millions of people. It will mean migration off the land at levels we have not seen before, and at levels poor countries cannot cope with."

Andrew Simms of the New Economics Foundation, one of Britain's leading experts on the effects of climate change on developing countries, remarked, "There's almost no aspect of life in the developing countries that these predictions don't undermine—the ability to grow food, the ability to have a safe sanitation system, the availability of water. For hundreds of millions of people for whom getting through the day is already a struggle, this is going to push them over the precipice."

A NOAA measure of drought from a climate model called the Palmer Drought Severity Index (PDSI), developed in the 1960s by Wayne Palmer, predicts there will be a notable increase in drought globally during this century with predicted charges in rainfall and heat around the world because of global warming. The PDSI figure for moderate drought is currently at 25 percent of the Earth's surface. By the year 2100, it predicts it will rise to 50 percent. The figure for severe drought is currently at 8 percent; by 2100 it is predicted to be 40 percent. The figure for extreme drought is currently at 3 percent; by 2100 it is predicted to rise to 30 percent, based on the study conducted by the Hadley Centre.

A response to this model result by Mark Lynas, author of *High Tide*, a book about global warming, was, "We're talking about 30 percent of the world's land surface becoming essentially uninhabitable in terms of agricultural production in the space of a few decades. These are parts of the world where hundreds of millions of people will no longer be able to feed themselves."

For a realistic glimpse into the future, predictions are that valleys once fertile will turn dry and brown. For inhabitants of desert valleys that rely on a short rainy season each year in order to be able to grow crops and graze their animals, instead they will wait for rains that will never come. Year after year, the situation will repeat itself as millions of people near starvation. Nomadic herders' animals will die; their cattle are emaciated to skin and bones now. Bleached skeletons of cows, goats, and sheep will litter the barren, dusty landscape. Nomadic herders will set out and walk for weeks without finding a watering hole or a riverbed. As people begin dying of starvation, inhabitants of different geographic areas will start fighting for what slim, meager resources still exist.

As brutal as this may sound, if global warming continues, this scenario is projected to become commonplace. According to the Hadley Centre, the number of food emergencies in Africa each year has almost tripled since the 1980s. Across sub-Saharan Africa, one in three people today is undernourished.

A news report from the World Wildlife Fund (WWF) on January 13, 2003, from Sydney, Australia, said human-induced global warming was a key factor in the severity of the 2002 drought in Australia. The report, entitled "Global Warming Contributes to Australia's Worst Drought," warns that higher temperatures and drier conditions have created greater bush fire danger than previous droughts. The drought has also negatively affected their agricultural areas. Professor David Karoly, coauthor of the report, stated, "The higher temperatures experienced throughout Australia last year are part of a national warming trend over the past 50 years which cannot be explained by natural climate variability alone. Most of the warming is likely due to the increase in greenhouse gases in the atmosphere from human activity such as burning fossil fuels for electricity and transport and from land clearing."

Dr. Karoly believes this is the first drought in Australia where the impact of human-induced global warming can be directly observed.

According to Anna Reynolds, WWF's Australia Climate Change Campaign manager, "Global warming is a reality that is affecting the livelihoods of rural Australians." The low rainfall and higher evaporation have adversely affected agricultural productivity with lower crop production leading to lower export earnings for farmers.

According to Fox News, in a release issued in June 2006 the United Nations said, "The world's deserts are under threat as never before, with global warming making lack of water an even bigger problem for the parched regions; those areas are facing dramatic changes." The report said that most of the 12 desert regions whose future climate was studied faced a much drier future. Experts predicted that rainfall would decrease as much as 20 percent by the end of the century due to human-induced climate change.

Increasing drought will also lead to the possibility of more wildfires. Some models that simulate global warming predict drier summers at high northern latitudes. According to the IPCC, there are already documented decreases in precipitation in parts of Africa, the Caribbean, and tropical Asia. The 1999 drought in the eastern United States may be a glimpse of what future conditions may be like as the Earth's atmosphere heats up.

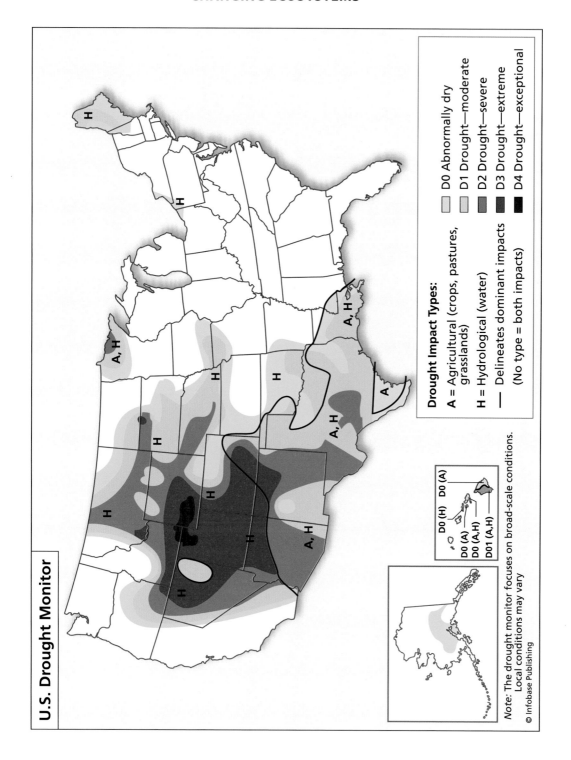

U.S. Drought Monitor

Drought Impact Types:

A = Agricultural (crops, pastures, grasslands)
H = Hydrological (water)
— Delineates dominant impacts
    (No type = both impacts)

D0 Abnormally dry
D1 Drought—moderate
D2 Drought—severe
D3 Drought—extreme
D4 Drought—exceptional

D0 (H)   D0 (A)
D0 (A)
D0 (A,H)
D01 (A,H)

Note: The drought monitor focuses on broad-scale conditions. Local conditions may vary

© Infobase Publishing

THREATENED DESERTS

In a report released in June 2006, the United Nations Environment Programme (UNEP) determined that the world's deserts are being threatened by global warming. The report, the "Global Deserts Outlook," warned that the world's deserts face dramatic changes because of global warming, such as high water demand and salt contamination of irrigated soils. Some of the most vulnerable locations within these regions are the desert margins and the mountainous areas within the deserts, such as the dry woodlands associated with desert mountain habitats, which may decline by up to 3.5 percent each year.

The Intergovernmental Panel on Climate Change (IPCC) has reported that the temperatures in deserts could rise by 5–6°F (3–4°C) by 2100. One of the worst impacts will be a dwindling water supply that currently depends on glacier melt as a water source, such as in central Asia, South America, and the southwestern United States. According to the IPCC, the glaciers on the Tibetan plateau may decline by as much as 80 percent by the end of this century.

As the atmospheric temperatures heat up, semiarid rangelands may be transformed into deserts. The mechanisms that could cause this degradation include water contamination from salinization and pollution by pesticides and herbicides. In addition, rising water tables beneath currently irrigated soils are also causing soils to have an elevated salt content, making them unfertile. In semiarid locations near oceans, the groundwater has been contaminated by seawater.

Nick Nuttall, a spokesman for UNEP, said, "Deserts are the last great wildernesses and the Cinderellas of the conservation world—out of sight, out of mind. Everybody cares about the mountains. Everybody is worried about the oceans . . . But nobody has really thought about the deserts before. They need help."

(opposite page) This map depicts the drought conditions for the western and central United States during May 2003. *(Modeled after R. Heim/C. Tankersley, NOAA's National Climatic Data Center)*

Based on UNEP's findings, it is predicted that rainfall could decrease by as much as 20 percent by the end of the century due to human-induced climate change.

In one global warming model, reported by National Geographic News in June 2005, it was suggested that global warming could stir up southern Africa's huge dune fields, letting loose great sand seas. Today, the Kalahari dune fields are stable because they are covered by vegetation. If droughtlike conditions take over and there is less rainfall and higher temperatures, the vegetation will dry up and die, no longer keeping the dune fields stable. If wind activity increases, the now exposed sand will be reactivated and cause problems for farmers.

If these massive dunes begin to shift, they will transform vegetated land into desert. The Kalahari region is enormous—it covers 1 million square miles (2.5 million km²) and stretches from northern South Africa to Zambia. In order to study possible dune behavior under global warming conditions, David Thomas, a physical geographer at Oxford University in England, used what he calls a "Dune Mobility Index" in a computer model. The Index looked both at erodibility (how easily dunes are able to erode) and erosivity (how much erosive energy the wind contains). The study found that the southern dune fields of Botswana and Namibia will become activated (erodible) by 2040. The northerly and easterly dunes in Zimbabwe and Zambia will begin to shift by 2070. By 2100, all the dunes in the Kalahari region are likely to be mobile.

According to Thomas, "These are areas where dunes are currently wooded in many places. We'll potentially see major environmental changes with currently vegetated, but sandy, landscapes reverting to active, blowing sand seas where life will potentially be very difficult."

Conditions like these predicted have not occurred in the Kalahari in more than 14,000 years. The sand would not constantly blow, however. It would blow during consistent time periods. Today, ranchers that graze livestock use many of these areas. Active dunes would ruin their herds' ability to graze, inflicting catastrophic consequences to their traditional livelihoods.

Nicholas Lancaster of the Division of Earth and Ecosystem Sciences at the Desert Research Institute in Reno, Nevada, remarked, "This is a very important study that shows how currently semiarid areas may respond to global warming. The implications for southern Africa are huge—especially for cattle herders, wildlife, and tourism."

Deserts are viewed as an especially critical ecosystem to protect and manage in a healthy way. Not only are they important for all the people who call them home, but these regions are also being considered as an enormous potential source for future energy. If the world's vast desert regions could be used as a receptacle of solar power, they could offer a feasible, environmentally friendly replacement for fossil fuel use. Some estimates say that an area of the Sahara 500 by 500 miles (800 by 800 km) could capture enough solar energy to meet the entire world's electricity needs. Similar to the world's rich rain forest ecosystems, deserts are also seen as sources for new drugs and crops.

An extremely interesting, relevant global warming study is being conducted in Karoo, a desert ecosystem in South Africa, considered to be one of the world's 25 biodiversity hot spots. Conservation International has a 45,000-square-mile (116,000-km²) site there that hosts about 5,000 different plants.

Guy Midgley, a plant physiologist at the National Botanical Institute (NBI) in Capetown, South Africa, is involved in determining the effects that rising temperatures have on the plants. Midgley and his colleagues have been working in Karoo since 2001, tracking signs of plant stress. And signs of stress have been plentiful. "With some plants, like proteas, we have seen areas of local extinction—dried husks of adult plants and no young plants. In other areas, shriveled leaves and aborted flowers reveal the lack of water and nutrients. Without flowers, there are no seeds; without seeds, no future generation of plants. In some areas, only adult plants remain. Young plants are not thriving because they don't have the water storage capacity to survive the heat. The adult plants are like the living dead," Midgley says. He is also looking at vegetation migration issues.

Brett Orlando, a climate change adviser for the World Conservation Union in Gland, Switzerland, said, "Midgley's work is right on the cutting edge. He is one of the first to look at an entire ecosystem.

One thing Midgley is doing is conducting experiments to see how well plants survive in increasingly warm conditions. He has set up 20 hexagonal Plexiglas chambers that trap the Sun's energy and raise the temperature around the plants between 4–6°F (2.4–3.6°C) during the day.

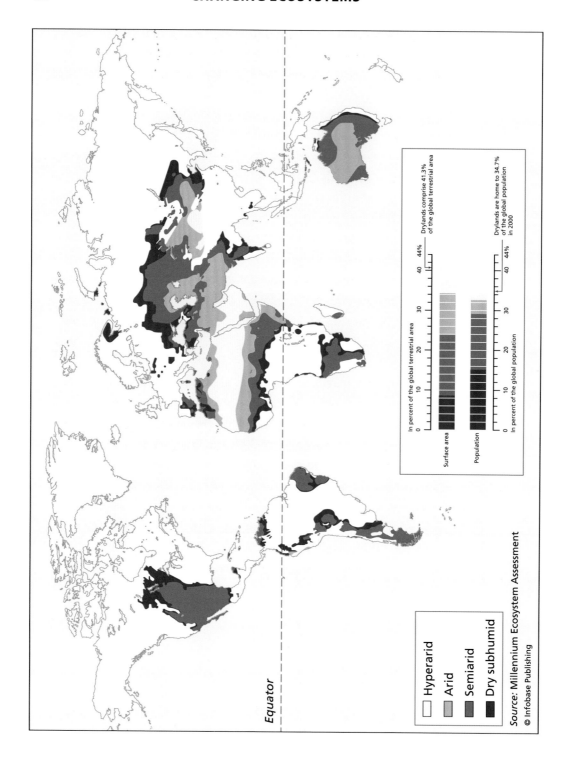

Hyperarid
Arid
Semiarid
Dry subhumid

Equator

Source: Millennium Ecosystem Assessment

© Infobase Publishing

In percent of the global terrestrial area

0   10   20   30   40   44%

Drylands comprise 41.3% of the global terrestrial area

In percent of the global population

0   10   20   30   40   44%

Drylands are home to 34.7% of the global population in 2000

Surface area

Population

This greenhouse effect kills 70 to 80 percent of them. "The models predict that the plants won't survive, and they don't," Midgley says.

Next, they want to test broader ranges of temperatures to see just how much the plants can tolerate before they begin to die.

"We want to know what is considered safe climate change," Midgley says.

Midgley believes that if these limits are known, it could set a precedent for global energy policies and land use, as well as encourage people to rely less on fossil fuels. It can also lead to better land management practices with the establishment of wildlife corridors.

DESERTIFICATION

One of the impacts that global warming will have is to exacerbate the worldwide progression of desertification. If there is a significant decrease in the amount of rainfall reaching an arid or semiarid area it could increase the extent of the dryland areas, destroying both vegetation and soils. Desertification is a degradation process inflicted on arid and semiarid landscapes due to human activities, climate change, or a combination of both. Although desertification has always existed, it has accelerated in recent years as populations have rapidly expanded across previously untouched landscapes and arable land has been cultivated and grazed. This had led to grave concerns.

Desertification first became well known in the United States in the 1930s during the dust bowl era, due to a combination of drought, improper farming, and land management practices. During this time, millions of people were forced from their homes in the biggest migration in U.S. history. Scores of people were forced to abandon their farms, leaving everything behind. The Great Plains were eventually restored over time through the practice of good land management plans, improved agricultural methods, and responsible conservation efforts.

(opposite page) This map shows where all the Earth's current drylands are located. They are some of the most vulnerable places on the globe. *(Source: Millennium Ecosystem Assessment)*

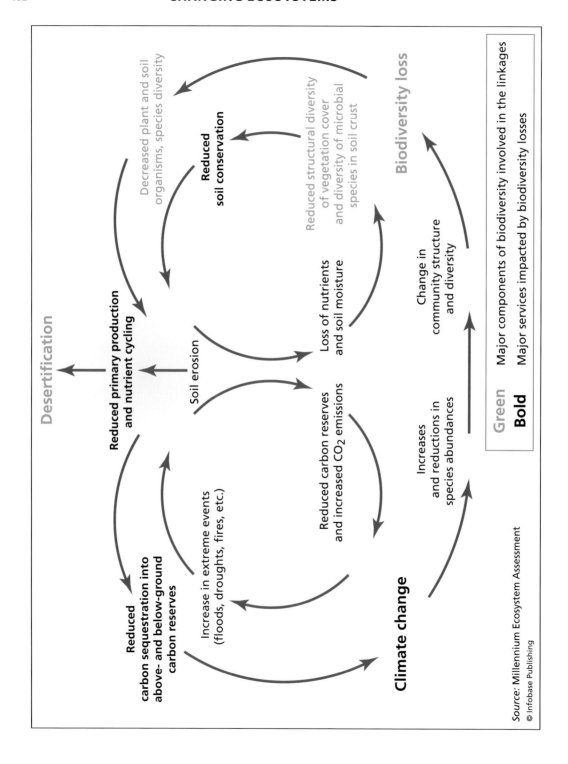

Desertification

Reduced primary production
and nutrient cycling

Decreased plant and soil
organisms, species diversity

Reduced
soil conservation

Reduced structural diversity
of vegetation cover
and diversity of microbial
species in soil crust

Biodiversity loss

Soil erosion

Loss of nutrients
and soil moisture

Change in
community structure
and diversity

Reduced
carbon sequestration into
above- and below-ground
carbon reserves

Increase in extreme events
(floods, droughts, fires, etc.)

Reduced carbon reserves
and increased CO_2 emissions

Increases
and reductions in
species abundances

Climate change

Green   Major components of biodiversity involved in the linkages
Bold   Major services impacted by biodiversity losses

Source: Millennium Ecosystem Assessment

© Infobase Publishing

In other areas of the world, however, population explosions and increasing livestock pressure on marginally healthy land have worsened the problem of desertification and sped it up. As desertification intensifies, the productive capability of the land decreases, destroying the original biodiversity. Different plant species produce physically and chemically different litter compositions in the ground. The litter, along with the natural biologic decomposers in the ground, helps form the soil complex and plays an active role in nutrient cycling. All the vegetation supports the primary production that provides food and wood that works together to sequester carbon, which plays an ultimate role in global climate. When these connections are broken, desertification is triggered and habitats can be lost. Biodiversity loss affects the health of the habitat. Global warming increases evapotranspiration and adversely affects biodiversity.

The loss of vegetation in the food chain affects all the life along it. One major problem is that when native vegetation dies off, invasive species, such as cheatgrass (*Bromus tectorums*) in the southwestern United States, moves in and takes over. Invasive species then have a better probability of survival because they do not have the predators in the environment that the native species do. It does not take long for invasive species to overrun a landscape, pushing native species out. Once this happens, it may be that only a few invasive species are supported in an area, where at one time dozens of species once existed. In areas where this occurs, the biodiversity is significantly lowered and unless rehabilitation efforts are put into effect, it may never be able to reestablish itself to a natural, ecologically balanced state. Even if native vegetation is reintroduced, it may not be able to survive if desertification degrades the land to the point where nutrients and water supplies are not available.

(opposite page) This flowchart represents the association of climate change, biodiversity loss, and desertification and shows how they work together in order to arrive at a better understanding of land degradation and global warming. *(Source: Millennium Ecosystem Assessment)*

Biodiversity, which contributes to many of the benefits provided to humans by dryland ecosystems, is greatly diminished by desertification. For example, diversity of vegetation is critical in soil conservation and in the regulation of surface water and local climate. If these delicate balances are disrupted, it can threaten the health and existence of habitats. Desertification affects global climate change through both soil and vegetation losses.

Another problem that exists involves the use of land for livestock grazing. When desertification becomes an issue and invasive weeds move into the area, the livestock may consider the new vegetation species unpalatable and refuse to graze.

It is not just one human activity that is responsible for desertification—there are many. Overgrazing of livestock and cultivation are two of the most common reasons, but deforestation is also a major contributor. If groundwater is overused or irrigation of farming areas is not done properly, it can introduce the desertification process. Desertification is also encouraged if soils collect more salt, raising their salinity levels. Global warming is also being blamed for worldwide desertification in the Earth's arid, semiarid, and subhumid areas. This means that desertification is the result of a combination of social, economic, political, physical, and natural factors, which vary from region to region.

Currently, lands that seem to be most prone to desertification include the areas at the fringes of deserts. These transition areas have fragile, delicately balanced ecosystems, usually operating under various microclimates. Already delicate, once these ecosystems are stressed to their limit, they cannot recover on their own. Grazing by livestock is especially harmful. As cattle and other animals graze, they pack down the soil with their hooves. The more the subsoil gets packed into an impermeable layer, the less water is able to percolate down into the ground. Because the scarce water is not able to penetrate the ground's subsurface it flows off the surface, eroding the land with rills and gullies. South Africa, for example, loses 300–400 million tons (262 million metric tons) of topsoil each year. In addition, as the surface dries out, the soil gets blown away in the wind.

According to the United States Geological Survey (USGS), there are many things scientists still do not know about desertification of productive lands and the expansion of deserts. To date, there is "no consensus among researchers as to the specific causes, extent, or degree of desertification. Desertification is a subtle and complex process."

In a global context, during the past 25 years, satellites have helped scientists study global warming by providing global satellite imagery for scientists to study the effects of desertification. The existence of satellite imagery has made it possible to monitor areas over time and determine the susceptibility of the land to desertification. Satellite data, such as that from Landsat, SPOT, Quickbird, and Digital Globe, have allowed scientists to accurately quantify and monitor the impacts of people and animals on the Earth.

The problem is a global one. It is predicted that by 2100 global warming will increase the area of desert climates by 17 percent, placing more areas at risk of desertification. Worldwide about 30 million acres (12 million ha) become useless for cultivation each year (this area equals about 87 percent of the agricultural land in the United States).

Desertification needs to be monitored and managed as a worldwide effort as global warming intensifies. Like global uplifting, it does not stop at international borders. If it is not controlled, biodiversity will be negatively affected. Another way desertification contributes to global warming is by allowing the carbon that has been stored in dryland vegetation and soils to be released to the atmosphere as it dries out and dies. This could have significant consequences for global climate. According to Greenfact's Desertification Synthesis Report, "Millennium Ecosystem Assessment in 2005," it is estimated that every year 300 million tons (272 million metric tons) of carbon are lost to the atmosphere as a result of desertification. This equates to around 4 percent of global emissions from all sources combined.

The relationship between global warming and desertification is not straightforward; its variabilities are extremely complex. As an illustration, the following are two potential scenarios:

1. Higher carbon dioxide (CO_2) levels in the air result in higher atmospheric temperatures. This dries the soil out, causing vegetation to dehydrate and die. The soils lose organic richness, and evapotranspiration becomes virtually nonexistent.

This creates a smaller carbon source over time. Because of this, the air is extremely dry, keeping water vapor from condensing into clouds. Desertlike conditions persist, and fringe areas become desertified.

2. In other areas, an increase in CO_2 may encourage an acceleration in plant growth and productivity. But if the ecosystem has become degraded to the point where invasive species have overtaken native species, even if native species are reintroduced into the ecosystem, the soil may be already degraded to the point where it will no longer survive due to lack of proper type and amounts of nutrients. In addition, if the Earth's energy or water balances are altered, this may also keep an original ecosystem from surviving.

The best way to deal with desertification is to prevent it from happening in the first place. The most appropriate way to do this is through proper management at local, regional, and global levels. Besides being environmentally better, prevention is also more cost effective than rehabilitation. The best form of prevention requires a change in the attitudes of those living on, and working with, the land. If sustainable, environmentally friendly agricultural and grazing practices are put into effect before the land becomes degraded, serious degradation problems can be prevented. For areas that have already been degraded, rehabilitation and restoration measures can help to restore lost ecosystems, habitat, and services the drylands originally supplied. One of the most important reasons to avoid desertification is to avoid extreme poverty and hunger—when drylands become too degraded in developing countries, populations are left with little access to food and clean water.

There are sustainable practices that can be implemented so that desertification is prevented. Integrated land and water management are key methods of desertification prevention. All measures that protect soils from erosion, salinization, and other forms of degradation effectively prevent desertification. In fact, dryland populations that build on long-term experience and active intervention can keep ahead of desertification by improving agriculture practices and enhancing pastoral mobility in a sustainable way. This productivity is already being seen in the Sahel region.

As global warming continues, these actions need to be taken to ensure that landscapes remain healthy and productive. According to the Ecosystems and Human Well-being: Desertification Synthesis, a report published in 2005 by the Millennium Ecosystem Assessment, the following actions can be employed to prevent desertification:

- implementation of a land and water management plan to protect soils from erosion, salinization, and degradation
- creation of economic opportunities outside of dryland areas, taking the stress off of drylands
- protection of vegetative cover so that it will stabilize the soil underneath and keep it from being eroded by wind and water
- becoming involved in alternative livelihoods that do not depend on intensive land use; such as greenhouse agriculture, tourism, and aquaculture
- combining areas of farming and grazing in order to centrally manage natural resources more effectively
- empowering local communities to effectively manage their own resources and combining traditional practices with local ones

In areas where desertification has already become established, it is important to rehabilitate and restore the lands in order for them to return to their previous conditions. Successful restoration must be done at the local level. Several methods are commonly used, such as the following:

- reintroduction of the original, natural species that used to live there
- combating erosion through the systematic terracing of steep areas so that water does not run down slopes, eroding the land's surface
- establishing seed banks to ensure that species do not become endangered or extinct, so that when climate conditions exist for the plant to survive, seeds can be planted
- enriching the soils with nutrients, making them more fertile and conducive to vegetative growth
- planting more trees

THE STRAIGHT FACTS ABOUT DESERTIFICATION

The following facts reflect the realities of desertification facing today's world:

- Roughly 8.9 billion acres (3.6 billion ha) of the world's 12.8 billion acres (5.2 billion ha) of dryland used for agriculture have been degraded by erosional processes.
- One out of every six people is directly affected by desertification.
- Desertification has forced many farmers to abandon farming and look for jobs in the city.
- Each year 20,000 square miles (51,800 km²) of land are destroyed by desertification.
- Desertification affects almost 75 percent of the land in North America.
- Dust from deserts can be blown great distances into cities. Dust has been blown from Africa to Europe and the United States. During the dust bowl, it was blown from Oklahoma and out across the Atlantic Ocean.
- Desertification destroys the topsoil of an area, making it unable to grow crops, support livestock, or provide suitable habitat for humans.

Desertification must be dealt with in order to reduce the risk of frequent sand and dust storms, as well as flooding due to poor irrigation practices or inadequate drainage. If these things happen, loss of topsoil and important nutrients would ultimately lead to a loss of vegetative cover. The absence of vegetation means the loss of a potential carbon repository. Once desertification gets a big enough grip on an area, it can lead to regional shifts in climate, which could further enhance the greenhouse effect.

According to UNEP, desertification causes crop losses of approximately $42 billion a year. The Sahel countries have been experiencing droughtlike conditions for the past 35 years with up to 20 percent less

- Climate change can trigger desertification as well as poor land management practices.
- The destruction caused by desertification carries a hefty price tag. More than $40 billion per year in agricultural goods are lost, causing an increase in agricultural prices, having a negative impact on consumers.
- According to a UN study, roughly 30 percent of the Earth's land is affected by drought. Every day, approximately 33,000 people starve to death.
- Desertification makes the environment more likely to experience wildfires.

Desertification is a condition that can be stopped, but it usually is not brought to the public eye until it is well under way, making rehabilitation of the landscape much more difficult and expensive. If it were brought to the public's attention sooner and not allowed to reach a critical point, it would make recovery much more simple. One way to ensure the success of this is through outreach and public education. By keeping people informed and educated about the results of their behavior on the environment, land degradation can be avoided completely.

rainfall. Semidesert regions have already advanced 60 miles (100 km) southward. Continental interior locations are expected to become drier.

According to the IPCC, many deserts will face a 5–15 percent decrease in rainfall. The Great Basin region of the southwestern United States is expected to be one of the hardest-hit drylands. In order to fight against desertification, action must be taken now. Results may take 10 or more years. Low-intensity agriculture will have to be controlled in recovery areas. Not only do the scientific and technical communities need to stand behind positive action, but leaders of today and tomorrow do as well.

In a study featured in the *Independent* in 2006, it was suggested that the Amazon rain forest could become a desert, which would then speed up the process of global warming with "incalculable consequences."

In studies conducted in the Amazon by the Woods Hole Research Center, it was found that the rain forest cannot withstand more than two consecutive years of drought without extreme ecological impacts and destruction. The research pointed to the fact that drought in the Amazon would trigger drought in the Northern Hemisphere and could accelerate global warming, creating a process that could get so far out of control that it could make the Earth uninhabitable.

In the rain forest the trees managed drought fairly well after the first year. Then, in the second year, they sunk their roots deeper in the ground to find moisture and still survived. But by the third year the trees began dying. The tallest trees were affected first. As they toppled to the forest floor, the lower canopies were exposed to direct sunlight. The forest floor was also exposed to direct sunlight and dried out quickly.

After the end of the third year, the biomass (vegetation) had released more than two-thirds of the stored CO_2 to the atmosphere. Where once the forests operated as a carbon sink, they were now a major carbon source.

What concerns scientists most is that the Amazon currently holds about 90 billion tons (81.6 billion metric tons) of carbon, enough to increase the rate of global warming by 50 percent. On top of that, if a wildfire were to start in these remote locations and destroy the vegetation, the rain forests would be transformed into a desert.

Dr. Deborah Clark from the University of Missouri, a renowned forest ecologist, says that research shows that "the lock has broken on the Amazon system. The Amazon is headed in a terrible direction."

HEAT WAVES

Temperatures are obtained from monitoring stations worldwide in order to calculate global mean temperature rise. When temperatures are taken near cities, they must be corrected to eliminate the specific effect that the presence of the urban area has on the temperature reading. Because urban areas have so many dark surfaces, such as asphalt-covered roads and dark roofs on buildings, they absorb more heat than natural surfaces such as grass, prairie, and woodlands. This absorbed heat is reradiated by the buildings and roads, and the resultant increase in temperature from these sources, as well as heat released from industry, cars, and other sources of burning fossil fuels, adds to the increased temperatures.

In order to use a reliable temperature value instead of a skewed value due to artificial inputs, collectively referred to as the urban heat island effect, this contribution to the temperature must be accounted for and removed. In addition, the various instruments and methods used worldwide must also be calibrated so that the temperatures collected are comparable.

According to UCS, with all of these conditions taken into account, the global mean temperature has increased approximately 0.5–1°F (0.3–0.6°C) over the past 150 years. More important, since 1975 the increase of the five-year mean temperature has been calculated at 0.8°F (0.5°C), which is a rate that is faster than any other five-year period in the past 150 years. Fourteen of the Earth's warmest recorded years have occurred since 1990. In addition, reconstruction of past climates through the use of *proxies* (items that present evidence of past climatic conditions) such as fossil pollen, tree rings, coral, ice cores, and sediment cores reveal that 20th-century warming is significantly different than in the past 400 to 600 years.

The only reliable explanation scientists have been able to come up with is human interference. The temperature rises in the models accurately reflect the actual temperature rises the Earth has experienced when the effects of greenhouse gas levels from the burning of fossil fuels (oil, gas, and coal) and deforestation are entered into the mathematical model equations. If the human interference factor is not added, the models do not work—they underestimate actual temperatures. The high latitudes (polar) have been identified in models as those areas on Earth that are being affected the fastest and the most significantly. In addition, nighttime temperatures have increased much more than daytime temperatures, keeping the Earth's atmosphere warmer overall. This is significant because when the Earth stays warmer at night, it retains the heat that was generated during the day in the atmosphere, starting the next day off warmer than normal.

The UCS believes that by the year 2100 there will be an increase in average surface air temperature of around 4.2°F (2.5°C). This will mean an increase in temperature at many scales, such as days, seasons, and years. When looking at significant local levels of temperature increases, this means that some areas will succumb to more extremely hot summer days during the summer and killer heat waves will occur.

According to A. Kattenberg, a climate adviser at the Royal Netherlands Meteorological Institute, temperate climates (such as the United States) would experience a doubling of extremely hot days, resulting in a 3.3–5°F (2–3°C) increase in average summer temperatures. D. J. Gaffenard and R. J. Ross of the R. J. Ross Company have researched increased summer heat stress in the United States and have determined that the number of days per year that temperature thresholds for death have been passed have increased significantly over the past 50 years for many U.S. cities.

As heat wave incidents increase, more people would be negatively affected and many of them could die. As an illustration, in 1995 in Chicago almost 500 people died during a significant heat wave. The sick, very young, elderly, and those who cannot afford indoor air-conditioning are the most at risk of dying.

In July and August 1999, another heat wave hit Chicago. Because emergency response network specialists learned some valuable lessons in 1995, they were better prepared for this one. Even so, 103 heat-related deaths still occurred.

Chicago is not alone. In 1980 a heat wave killed 1,700 people in the East and Midwest, and in 1988 one killed 454 people. In 1998, more than 120 people died in a Texas heat wave.

Europe experienced a deadly heat wave in 2003, so hot in fact that it was considered the hottest summer in 500 years. During this period, 27,000 people died from heat-related problems. Some of those who did not die suffered irreversible brain damage from advanced fevers as a result of the intense temperatures.

The heat wave in the United States in 2006 was one of the worst it had ever experienced. This one held the entire country in its grip and lasted for almost a month. The effects and costs were enormous—hundreds of people died, massive power outages were triggered, and unmanageable wildfires burned large areas. According to the Environmental Defense Fund (EDF), tens of thousands of people in New York went without electricity for more than a week.

One thing that scientists cannot do with the overall issue of global warming is to blame a single weather event—like a heat wave, a hurricane, a blizzard, or a tornado—on global warming (because weather fluctuates naturally). But what they can relate to global warming are trends. Based on the results of climate models, trends of more wild and unpredictable weather have been forecast as one of the results of global warming, and an increase in heat waves does fit into this future climate scenario. One thing that both developed and developing countries need to be aware of is that continued urbanization will increase the number of people vulnerable to these urban heat islands and heat waves.

WILDFIRES

The likelihood of a disastrous wildfire occurring increases significantly during periods of drought. The world saw proof of this in 1997 and 1998 when an El Niño episode caused extremely dry conditions in many areas across the globe. Large forest fires occurred in Brazil, Central America, Florida, eastern Russia, and Indonesia. Wildfires can easily occur during drought periods from both natural and human-caused factors. Lightning strikes are a common cause of them, as well as campers or hikers not practicing proper campfire control behavior. Regardless of the cause, during drought conditions, the vegetation is so dry that it does not take much to start a wildfire, and when they start, they burn fast and intensely. They are often difficult to extinguish and put thousands of firefighters in grave danger.

According to a report on CNN on November 2, 2000, global warming may be a significant contributor to accelerating the fire cycle in desert ecosystems of North America. Quoting a report that was published in the journal *Nature,* CNN stated that rising CO_2 levels from burning fossil fuels can negatively alter the delicate desert ecosystem.

Stan Smith, a professor of biology at the University of Nevada in Las Vegas, as well as lead author of the study, said, "This could be a real problem for land managers."

Smith and his team of scientists ran an experiment in the desert where they increased the CO_2 emissions by 50 percent and then monitored the resultant impact on four plant communities in the Mojave Desert ecosystem. With the elevated CO_2 levels, all the plants' density and biomass increased. The scientists likened this result to what would happen in nature during high precipitation years with increased CO_2. Then, when drought occurred during an El Niño, this extensive biomass became dried out tinder, highly susceptible to wildfire.

Smith noted, "Recent studies suggest that El Niño high rain cycles will intensify with climate change."

In addition, the increased CO_2 levels also encouraged the growth of invasive plants. One in particular—red brome—grew abundantly in the study area. This invasive species not only did well in the ecosystem, but it also contributed to a high load of burning as a result of the drought.

"Red brome is capable of carrying fire across the bare zones between shrubs, and thus burn both the grasses and shrubs. Prior to the introduction of these exotic grasses, there were few plant species that could create such a continuous cover, and so these desert scrub ecosystems did not historically burn," Smith reported.

This situation is not only bad news in terms of creating an enhanced wildfire cycle, it is devastating news in terms of ecosystem health and severe loss of biodiversity. Conservation groups, such as the Nature Conservancy, cite climate change as one of the causes behind the 2000 fire season, which charred more than 5 million acres (2 million ha) in the western United States.

Bob Nowak, coauthor of the study, said, "In a lot of ways the experiment we're doing is a look into the future, and the future doesn't look so good. What we are already seeing in the northern Nevada Great Basin with cheatgrass, we are going to see more and more of in southern Nevada in the Mojave with red brome."

Based on a report from the Natural Resources Defense Council (NRDC), the 2006 wildland fire season set new records in both the number of reported fires as well as the acres burned. Close to 100,000 fires were reported and nearly 10 million acres (4 million ha) burned—125 percent above the 10-year average. If warming continues to spur wildfire seasons, it could be economically devastating. Firefighting expenditures recently have totaled $1 billion per year.

SOUTHERN CALIFORNIA ON FIRE

Severe wildfires threatened the lives and property of residents of Southern California in 2007 until the fires were finally extinguished. The following are some of the facts about one of the worst five years in California's history:

- Fires that erupted in San Diego forced more than 265,000 people from their homes.
- There were approximately 4,900 firefighters involved in battling the blazes.
- Governor Arnold Schwarzenegger declared a state of emergency in seven counties and requested that the National Guard pull 800 soldiers from patrolling the United States–Mexico border to help fight the fires. He also requested the Defense Secretary Robert Gates to supply all of the military's available Modular Airborne Fire Fighting Systems (MAFFS) to help fight the fires.
- The Department of Agriculture provided crews, tankers, and helicopters to help battle the fires.
- The San Diego region was hit especially hard because it had previously experienced heat waves, very dry landscapes, and extremely powerful winds.
- Qualcomm Stadium, where the San Diego Chargers play, was converted into an evacuation facility.
- There were times when the fixed-wing firefighting aircraft were grounded by strong winds, unable to fly. This made the job even more difficult for the ground-based firefighters.
- Fires threatened the San Diego Zoo's Wild Animal Park. The park had to close, and some of its endangered species, such as the condors, were moved to a safer location until the fires were under control.
- People were asked not to call 911 unless it was a dire emergency because the cell phone lines were clogged and rescue attempts were being hampered.
- The largest of the California wildfires was the Buckweed blaze north of Los Angeles. It scorched 27,500 acres (11,129 ha) and forced 15,000 people to evacuate their homes.

As a result of the 2007 wildfires, nearly 39,000 homeowner insurance claims were filed and $13 million in losses were paid by insurance companies. The fires caused $2.3 billion in losses.

In 2007, California was subjected to a series of devastating wildfires that began burning across Southern California in October. More than 1,500 homes were destroyed and more than 500,000 acres (202,000 ha) of land were burned from Santa Barbara County southward to the United States–Mexico border. Nine people died and 85 were injured.

These fires occurred at the end of a very dry summer and were made even worse by the Santa Ana winds, which were blowing 60 miles per hour (97 km/hr), fueling the fire. In addition, temperatures were consistently in the 90°F (32°C) range, setting the stage for destructive, dangerous wildfires to do the most damage.

When the last of the fires were extinguished, the number of displaced residents who had been evacuated totaled above 900,000. It was the largest evacuation in California's history.

Air condition and quality also suffered because of the concentrations of particulate matter, 10 micrometers (0.0004 inch) and smaller (referred to as PM10), which reached unhealthy levels in the atmosphere, affecting those with breathing problems. It reached such unsafe levels that residents were requested to consider a voluntary evacuation from the city until the particulates were cleared from the air.

As global warming continues, these issues are expected to increase in frequency and duration. Drought, desertification, and wildfires are a combination expected to become more of a problem as global warming accelerates.

Impacts to Mountain Ecosystems

Mountain habitats worldwide provide some of the most inspiring scenery on Earth. Abundant vegetation and wildlife draw millions of visitors each year to enjoy such activities as camping, hiking, wildlife watching, canoeing, mountain climbing, skiing, or just relaxing and enjoying the clear air and pristine views. However, with the steady increase in temperature, mountain ecosystems are beginning to pay the price. These regions, compared with even a decade ago, are in a state of transition—and some are already gone. The mountain ecosystem has a highly specialized niche, and even small changes can cause species to become threatened, endangered, and extinct.

This chapter explores the various ways in which global warming is changing the world's mountain habitats and the dangers that now exist because of this change. It then looks at some economic challenges and presents a case study to illustrate how much impact a few degrees can really have on the health and future of a mountain ecosystem.

MOUNTAIN ECOSYSTEMS IN DANGER WORLDWIDE

According to National Geographic News on February 1, 2002, the United Nations University in Tokyo has determined that global warming, in addition to other issues such as pollution, deforestation, and overuse and development, are having a huge impact on mountain environments today and stressing them beyond the point of being able to remain healthy.

According to Dr. Jack Ives, a professor of geography and environmental science at Carleton University in Ottawa, Canada, as these mountain systems become more taxed, "water shortages, landslides, avalanches, and catastrophic flooding will become major consequences." Water shortages alone present a critical issue. Mountainous regions occupy roughly 25 percent of the Earth's land surface and provide a home to 10 percent of the world's population but also function as the principal

Montane and mountain environments exist at high elevations and have distinctive vegetation zones. As species begin to migrate upslope under hotter temperatures, the vegetation originally in the upper elevations will be pushed out of the environment completely with no place left to go. *(Nature's Images)*

source of water for half of the world's population. If global warming has a negative impact on mountain ecosystems, worldwide disaster would result.

According to Dr. Ives, "The most severe examples of environmental and socioeconomic degradation—now near total disaster—are the Hindu Kush in Afghanistan, the Karakorum and western Himalaya, and the disputed territory of Kashmir."

Other mountain ranges in trouble included in the United Nations' report as "ecologically endangered" are the European Alps, the Sierra Chincua in central Mexico, the Amber Mountains in Madagascar, the Snowy Mountains in Australia, and the Rocky Mountains in North America.

The problem consistent with all the mountain ranges is the melting of their glaciers and ice caps. According to the World Glacier Monitoring Service in Switzerland, the world's glaciers have been shrinking faster than they have been growing—there is a net loss worldwide because of global warming. They identified the losses during 1997–98 as some of the most "extreme" and predict that up to 25 percent of the Earth's glacier mass could completely melt by 2050 and up to 50 percent by 2100. If this were to occur, the only areas left on Earth that would have any remaining glacial deposits would be Patagonia, the Himalayas, and Alaska.

The United Nations reported that Mount Kilimanjaro in Tanzania has lost 82 percent of its mass over the last 100 years. In addition, the glaciers in Montana's Glacier National Park are melting so rapidly they could be completely gone in 30 years. Losing these ice-covered surfaces would have a negative impact on the Earth's energy balance. According to Lisa Mastny of Worldwatch Institute, "When the ice melts, newly exposed land and water surfaces retain heat, leading to even more melt and creating a feedback loop that accelerates the overall warming."

A serious consequence of these changing mountain ecosystems is the loss of species. As the climate warms, the unique plants and animals that cannot adapt will be forced to extinction. The United Nations reports that just in the Snowy Mountains of Australia the warmer temperatures and lack of snowfall of the past few years threaten more than 250 species of plants. Vegetation has already begun to migrate in response. Subalpine trees are now growing at altitudes 131 feet (40 m) higher than they were 25 years ago.

Another issue documented in the United Nations' report was that of insect infestations. Warmer temperatures encourage insects to migrate farther north into territories where it previously was too cold for them to survive. The Rocky Mountain chain of North America has experienced several insect infestations in the past decade in the Canadian Rockies. The consistent warm winters of the past few years have triggered a pine beetle infestation that has now diseased the trees of more than 1,931 square miles (5,000 km²) of the pristine forested areas of British Columbia. Mountain forest stands in Alaska are facing the same dilemma.

LACK OF WATER STORAGE

In a report issued by Space Daily in 2004, based on a recently developed climate change model, global warming is expected to reduce the amount of water stored as snow in the western United States by up to 70 percent over the next 50 years. A reduction in snowfall means much more than a shortage of drinking water for the next summer. If there is a reduction in the snowfall that covers the Sierra Nevada that feeds California and that which reaches the Cascades of the Pacific Northwest, it will lead to increased fall and winter flooding because the snowpack will not stay frozen. Without sufficient snowpack, there will not be enough water stored in the mountains to supply an adequate amount to residents along the populated West Coast. Wildlife indigenous to the region will also suffer. During droughts, water is often rationed and residents are restricted as to how much personal water they can consume for household use and landscaping. Significant amounts of water are also needed for industry and commercial businesses. All this must be budgeted and accounted for during years of water shortage. Some of the biggest impacts in the West will be felt at West Coast fisheries, agriculture, and hydropower generation.

L. Ruby Leung, a scientist at the Department of Energy's Pacific Northwest National Laboratory, said that this prediction concerning water storage along the West Coast is "a best case scenario."

The model developed jointly by the Department of Energy and the National Center for Atmospheric Research is on the conservative end

As global warming causes temperatures to rise, species will be forced to migrate to new environments. As plants and other food sources migrate northward and/or higher up mountain ecosystems, wildlife will have to follow. *(NOAA)*

compared with most climate prediction models. This model assumes a 1 percent increase each year in the rate of greenhouse gas concentrations through the year 2100, little change in precipitation, and an average temperature increase of 2.5–3.3°F (1.5–2°C) through 2050.

The model's result was that most of the precipitation would fall as rain instead of snow—two-tenths of an inch to more than half an inch each day, which would push the snowline in the mountains from 3,000 feet (914 m) to 4,000 feet (1,219 m) and higher. This would mean that instead of snow melting in late April, by 2050 the snow would melt much earlier. In fact, over the past 50 years, the coastal mountain ranges have lost 60 percent of their normal snowpack.

Leung says, "The change in the timing of the water flow is not welcome. The rules we have now for managing dams and reservoirs and irrigation schedules cannot mitigate for the negative effects of climate change."

Increasing temperatures will cause mountain environments to become drier, increasing the highly flammable tinder in forest areas. Both natural lightning strikes and human carelessness will increase the risk of forest fires. The drier conditions will make the fires spread more quickly and be harder to extinguish, putting both the natural environment and bordering urban areas at extreme risk and causing millions of dollars in damage. *(Kelly Rigby, Bureau of Land Management)*

Another issue that will make this more difficult to manage is future population growth. If the West Coast's population continues to expand, there will be an even higher demand on the water supply, and global warming and drought will exacerbate the problem further. Related to this problem is the issue of wildfire.

According to an August 2006 article in *Science* magazine, warmer temperatures are linked to the increasing duration and intensity of the wildfire season in the western United States. In fact, since 1986, the longer, warmer summers experienced in this region have resulted in a fourfold increase of major wildfires and a sixfold increase in the area of burned forest, as compared with between 1970 and 1986. Interestingly, a similar increase in wildfire incidences has also been reported in Canada from 1920 to 1999. Based on studies conducted by Anthony L. Westerling, a climatologist at the Sierra Nevada Research Institute, the length of the active wildfire season (the period when fires are actually burning) in the western United States has increased by 78 days and the average burn duration of large fires has increased from 7.5 to 37.1 days. He attributes this increase in wildfire activity to an increase in spring and summer temperatures by 1.5°F (0.9°C) and a one- to four-week earlier melting of mountain snowpacks. He says the snow-dominated forests at elevations of 6,890 feet (2,100 m) show the greatest increase in wildfire activity. Once the annual snowmelt is complete, the forests can become combustible within 30 days due to the low humidity and sparse summer rainfall typical of the region. Most of the wildfires in the western United States are caused by lightning and human carelessness. Because the forests are extremely dry and winds are usually present, the forests are in prime condition for burning and it does not take much to get a wildfire started. The extremely dry tinder present can make a fire run out of control very quickly. Often, these fires are in very remote locations, making it difficult for firefighters to get to them. In addition, these wildfires spread so rapidly that the growing populations adjacent to forested areas face huge risks of being burnt, as seen in the incredible destruction done in California during the wildfire season of the summer and fall of 2007.

Australia has also battled wildfires that have burned out of control. With its record highest temperature since 1939, Sydney experienced its hottest New Year's Day in 2006. Temperatures reached 112°F (44.2°C), causing power blackouts and igniting more than 40 bushfires along Australia's east coast. Nightmarish conditions played out as the flames roared uncontrollably. Homes and cars were destroyed, towns were isolated, and roads were blocked off. Flames 65 feet (20 m) high swept across coastal towns. Countless homes, buildings, farms, and other areas were burnt during this 22,000-acre (8,903-ha) wildfire.

Then, in February 2009, Australia experienced the deadliest wildfires in its history. People were burned in their homes and cars, and entire towns were obliterated. As searing temperatures and windblasts fueled the ranging fires, the flames ignited the drought-stricken country. Because of the forests' extra-dry conditions, these wildfires were the worst ever experienced in Australia. In Victoria, the countryside was completely blackened. Entire forests were reduced to leafless, charred trunks, and farmlands were reduced to ashes. The Victoria Country Fire Service estimated 850 square miles (2,200 km²) were completely burned.

Prime Minister Kevin Rudd said, "Hell in all its fury has visited the good people of Victoria. It's an appalling tragedy for the nation."

According to a study in *NewScientist,* Australia may have had, with the wildfires in February 2009, a horrifying preview of what climate change has in store for its people. They claim that climate models based on figures from the Intergovernmental Panel on Climate Change (IPCC) predict more frequent—and more extreme—fires for southern Australia over the next few decades.

According to John Handmer of the Bushfire Cooperative Research Centre at RMIT University in Melbourne, however, the role of climate change in recent fires has been downplayed. "Last weekend's fires were unprecedented. We had a record heat wave, the worst fire danger index on record, during a record-breaking drought."

The fire danger index takes into account both temperature and humidity. Over 50 is extreme; on Saturday, February 14, 2009, the index is believed to have been five to six times higher. As global warming continues, wildfires are expected to increase.

GLACIERS AND FLOODING

One of the prominent concerns in mountain ecosystems in light of global warming is the melting of glaciers and flooding of valley bottoms and other lowlands upon the rupture of huge meltwater lakes that have built up behind glaciers. As glaciers begin to melt and become unstable, it does not take much force from the water backed up behind them to push its way forward, escape its temporary dam, and rush downhill destroying everything in its path.

Martin Beniston, a climate scientist at Freiburg University in Switzerland, says, "In the Himalayas, some glaciers are up to 43 miles (70 km) long. In Bhutan alone, there are at least 50 lakes in this category, and a similar number in Nepal, as well. Towns and villages in their path could be hit by a tsunami."

One lake like this that formed in France's Savoie (Savoy) region reached the point where engineers were concerned the Rochemelon glacier that was holding it back would burst. They took corrective action by draining the lake before it could burst and cause a disaster. Glaciers in the Alps in Europe, the Andes in South America, and the Himalayas have been steadily retreating for the past five years.

Not only has the climate warmed up, but these areas have also received less snowfall. According to Heinz Slupetzky of the University of Utrecht in the Netherlands, all of the glaciers could be melted as early as 2080. In another part of the world, Yao Tandong, a glacier expert from China's Institute of Tibetan Plateau Research at the Chinese Academy of Sciences said that up to 64 percent of China's high-altitude glaciers could be completely melted by 2050. As a comparison to the volume of ice that represents, he compared it to an amount equivalent to all the water in the Yellow River. One of the most crippling side effects of this is that 23 percent of China's 1.3 billion people currently depend on the glacial runoff as their source of drinking water.

In 2002, a team of mountaineers, backed by the United Nations Environment Programme (UNEP), climbed Mount Everest in order to determine its current environmental health and the impact that global warming had had on the environment. The first thing the team documented was what they referred to as the "startling evidence of the impacts of climate change."

Roger Payne, who was one of the expedition's leaders, as well as a director at the International Mountaineering and Climbing Federation, remarked: "It is clear that global warming is emerging as one, if not the, biggest threats to mountain areas. The evidence of climate change was all around us, from huge scars gouged in the landscape by sudden, glacial floods to the lakes swollen by melting glaciers. But it is the observations of some of the people we met, many of whom have lived in the area all their lives, that really hit home."

Interviews with local people gave the mountaineers a unique insight into the rapid changes that have occurred in the area in recent years. According to Tashi Janghu Sherpa, president of the Nepal Mountain

Association, he has "seen rapid and significant changes over the past 20 years in the ice fields and these changes appeared to be accelerating." As a graphic comparison, since the time 50 years ago when Sir Edmund Hillary and Tenzing Norgay were the first to climb Mount Everest, the glacier they started their climb from has retreated three miles (five km). Today, climbers must walk two hours just to reach the edge of that same glacier.

Mr. Janghu also expressed concern over many of the areas on the glaciers that once contained small isolated ponds. Today, there has been so much glacial melting in the area that the network of small ponds has

A SPECIES THREATENED: THE ROCKY MOUNTAIN PIKA

The Rocky Mountain pika is a tailless relative of the rabbit and lives at high altitudes. A tiny animal full of energy, it can often be seen at the higher elevations darting from rock to rock in search of hay, building up its stash for the long, cold winter ahead.

Denise Dearing, associate professor of biology at the University of Utah, says, "They're a very charismatic animal. People like seeing them in the mountains. People go to the mountains oftentimes to see pikas and marmots, so it would be a shame to lose them." She is worried that pika will become one of the first mammals to fall victim to global warming.

To back this up, the United States Geological Survey (USGS) surveyed 25 populations around the American West, where pikas had been recorded. During the last century, where there was over a degree in temperature rise, pikas disappeared from roughly one-third of those states.

Pikas are very temperature sensitive. Because their fur is extremely dense, they cannot survive in warm temperatures. For example, they cannot even survive six hours in 77°F (25°C) heat. Pikas are especially at risk because their habitat is physically restricted to such a narrow range—located in the limited regions that fall between the tree line and the mountain peaks. The problem their habitat will face with global warming is that as temperatures rise, the pikas normally found at lower elevations on the mountain will begin to migrate upward to remain in contact merged into a mile-long lake. The locals are extremely worried that as glaciers continue melting the meltwater would cause huge floods and mudslides down the valleys. One glacial lake that had recently burst flooded a valley and washed away all the bridges downstream.

Scientists from UNEP have used satellite imagery to identify 44 glacial lakes in Nepal and Bhutan that are currently so swollen they could burst anytime within the next five years.

One of the guides for the expedition, Pemba Geljen Sherpa, said, "It is all changing—you do not see the same traditional dancing or singing of my parent's generation. But we need more, not less, tourism here

with the cooler temperatures they are accustomed to. As temperatures continue to rise, this will have a ripple effect, forcing species to continue adapting by moving higher in elevation. As a result, during this process the pikas will be forced all the way to the mountain peaks. At that point, there will be no place left for them to go, and when the mountain peaks become inhospitable to them, they will not be able to survive.

According to Dr. Dearing, the added carbon dioxide (CO_2) in the atmosphere as a result of global warming can also make the plants the pikas feed on toxic. She adds: "Pikas shelter under snow, but climate models suggest mountain snowpacks which provide 75 percent of the West's water are decreasing."

Dr. Dearing is not the only scientist in the West with these observations. Others are also making connections between rising temperatures, more extreme weather events, more and larger western wildfires, and more species on the brink of extinction.

Fred Wagner, a professor emeritus at Utah State University, remarked: "The way we live, things are going to get hotter. Our summers are going to get much warmer. Our winters are going to be milder. We're going to see snowpacks in the mountain ranges shrinking. That's happening over the West as a whole."

Dr. Dearing stresses, "Pikas are vocal animals and their silence is telling us we need to take immediate action against global warming."

to boost the economy and give people jobs, incomes, and education. What we cannot control is global warming; that is in the hands of others. We, here in Nepal, produce tiny amounts of the gases linked with global warming. It is up to the big, industrial countries of Europe, North America, and Japan to act to save our mountains and the environment upon which over livelihoods depend."

CHALLENGES IN ALPINE REGIONS

The mountain regions are experiencing some of the most profound changes due to global warming. In addition to the obvious loss of glaciers, which is easily quantifiable, hotter temperatures are driving the alpine zones farther toward the mountain summits. Alpine zones are usually found at an altitude of about 10,000 feet (3,048 m) or higher. They lie just below the snow line of a mountain. Before long, under the constantly rising temperatures, the alpine plants and animals will have nowhere else to go. Species that cannot adapt will not survive. This is already happening with the destruction of pika habitat in many portions of the Rocky Mountains today.

The vegetation is facing the same dilemma. On Mount Rainier and in Olympic National Park, mountain hemlock and subalpine firs have been documented moving into areas that have traditionally served as alpine meadows. In Yellowstone National Park, a similar migration is occurring with white bark pine. It is now migrating toward the mountain summits. If the white bark pine cannot find suitable habitat, it will affect the grizzly bear population because it is a significant food source for the grizzly. Global warming can jeopardize entire food webs.

In Europe, scientists worry that the biodiversity and survival of the fragile alpine life-forms, such as the delicate mosses and flowers, are in jeopardy. According to Georg Grabherr at the University of Vienna, if global warming continues, the alpine vegetation in Europe will migrate northward and eastward. He predicts that vegetation that is currently found in the Mediterranean will eventually be found in central Europe. Studies he has already conducted support his predictions—he has confirmed that alpine vegetation is already beginning to shift. Plants are not only migrating north, they are also moving upward in elevation on mountain ranges.

One of the big concerns he has expressed is whether or not vegetation will be able to migrate northward because so much of the vegetation's natural habitat has been taken for urbanization and other human uses. According to Grabherr, "When we talk about biodiversity, we have to consider special habitats which can disappear quite quickly. The summer drought of 2003 killed a lot of the grasses on the rocks where there is normally a lot of water. The purple mountain saxifrage could die from 'overheating' even though it is in the traditional 'cold climate of the Alps.'"

As stated by Vera Markgraf of the Institute of Arctic and Alpine Research at the University of Colorado at Boulder, "You reduce the capacity of plants to propagate if the habitat is already occupied or not available, as the situation is today where most habitats have been altered by human intervention."

ECONOMIC CHALLENGES

The mountain areas will also be vulnerable to and feel the effects of economic challenges. As forests succumb to drier conditions and suffer the damages of forest fires and invasive species degradation, the quality of their natural resources will begin to be challenged and dwindle. The effects of wildfire, insect and disease infestation, drought, and species migration will begin to change the health and biodiversity of the mountain ecosystems to the point where fishing, camping, hiking, backpacking, and boating may all be negatively affected in years to come.

According to the Outdoor Industry Foundation, in a study of the economic impacts on the industry released at the Outdoor Retailer Summer Market in Salt Lake City, Utah, in August 2006, the outdoor recreation industry has a $730 billion impact on the national economy. They reported that: "This amount factors in the amount Americans spend on outdoor trips and gear, the companies that provide that gear and related services, and the companies that support them. The outdoor industry also supports 6.5 million jobs, which amounts to one in 20 U.S. jobs, and generates about $88 billion in federal and state tax revenue while stimulating 8 percent of all consumer spending."

If global warming has a negative impact on the recreational uses of mountain areas, this will have a ripple effect throughout the local,

As global warming progresses, ecosystems will change, negatively
affecting the traditional recreational uses of mountain environments,
such as fishing. *(Nature's Images)*

national, and international economy, particularly in business sectors
such as the travel industry, the transportation industry, the sports equip-
ment industry, the food industry, and others related to leisure activities.

One specific industry in danger of surviving in a warming world is
the ski industry. According to a report in *USA Today,* global warming
"poses a serious threat" to alpine ski resorts and the regional economies
that depend on them, especially in Germany. In fact, the Organization for
Economic Cooperation and Development (OECD) studied the ski resorts
in the European Alps and reported that the mountain environment was
"particularly sensitive" to climate change. According to the report, "Recent
warming there has been roughly three times the global average."

Their studies showed that 1994, 2000, 2002, and 2003 were the
warmest years on record in 500 years. They also reported that "Climate
model projections show even greater changes in the coming decades,
with less snow at low altitudes and receding glaciers and melting permafrost higher up."

Coming at the same time as this report in December 2006 was the World Cup ski circuit, which was negatively affected by lack of snow—several of its events had to be canceled.

According to the OECD, a warming of 1.7°F (1°C) is enough to cause a 60 percent drop in the number of areas where there is reliable snow. The OECD noted that the low-altitude areas were more vulnerable than high-altitude regions. They also warn that "sufficient snow may become a thing of the past on lower lying slopes, especially in Germany, followed by France and Switzerland."

News of this prognosis has already reached the economic industry that supports the ski resorts. OECD official Shardal Agrawala reported that some of the smaller resorts were "already closing up."

One of the mountain recreation industries that is expected to be hit the hardest with the progression of global warming is skiing. The future of the world's ski resorts hangs in the balance as global temperatures rise. Pictured here is Ryan Hunter, a ski racer at the Super G event at Park City Mountain Resort in Utah, 2002. *(Nature's Images)*

The situation is serious enough that some banks in Switzerland are already refusing to lend money to ski resorts that are built in areas below an altitude of 4,921 feet (1,500 m). A current model's projections of warming say that by the years 2020–25, with a 1.7°F (1°C) increase in temperature, there may only be about 500 resorts in Europe even able to support any type of skiing activity. If there were to be a 3.3°F (2°C) rise, only 400 resorts could survive. This is a dire forecast considering that OECD says, "Tourism in the Alps is a key contributor to the economy of alpine countries. Between 60–80 million tourists go on winter holidays in the Alps each year."

Another report issued in December 2003, based on a study done by the University of Zurich for UNEP, stated that "Global warming could close half of the alpine ski resorts by 2050." They based their conclusions on estimates that temperatures could increase 3.3–9.7°F (1.4–5.8°C) by 2100. This would raise the snow line by 1,000 feet (305 m).

Identified in the study were Kitzbühel in Austria and Oberstdorf in Germany, which could receive so little snowfall over the next 30 to 50 years that skiing, snowboarding, and tobogganing could cease to be viable winter industries.

According to Dr. Rolf Burki, who led the study, "As ski resorts in lower altitudes face bankruptcy, so the pressure on highly environmentally sensitive upper-altitude areas rises, along with the pressures to build new ski lifts and other infrastructure."

Klaus Toepfer, who is the executive director for UNEP, stated, "Climate change in the form of extreme weather events such as hurricanes, floods, and droughts is the greatest challenge facing the world. But this study shows that it is not just the developing world that will suffer. Even rich nations are facing potentially massive upheavals with significant economic, social, and cultural implications."

CASE STUDY—WHITE MOUNTAINS, NEW ENGLAND

The effects of global warming have been felt in many areas and its specific components have been studied. This section looks at one area in particular—the White Mountains of New England. The White Mountains are located within the states of New Hampshire and Maine. Based on global climate models, warming there is predicted to increase and be

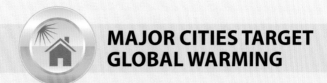

MAJOR CITIES TARGET GLOBAL WARMING

Denver, Colorado, has plans to become one of the nation's leaders in fighting global warming. Attacking the issue from several directions, Mayor John Hickenlooper has made their "climate action plan" a priority. It will focus on reducing gas emissions by moving to alternative energy sources, increasing recycling, and changing building codes to ensure energy conservation.

The plan does not just focus on those in the public who are willing to make these changes—the plan also has some bite to it. It includes penalizing heavy users of electricity and natural gas and basing auto insurance premiums on the number of miles traveled.

If this plan succeeds, Mayor Hickenlooper believes CO_2 emissions can be reduced by 4.8 million tons (4.4 million metric tons) by 2020—which is roughly the equivalent of eliminating two small coal-fired power plants or taking 500,000 cars off the road.

City officials have identified several ways to accomplish these ambitious goals:

- A large part of the plan involves finding ways to encourage energy conservation by mandating efficiency standards for new construction and setting standards for older homes that would be enforced when the home is sold.
- Incentives would be given for carpooling and the use of hybrids and other low-polluting vehicles, such as priority in parking.
- To reduce the problem of methane (a significant greenhouse gas) from landfills, the plan would encourage recycling and charge residents for the amount of trash they throw away each week.

Denver is patterning their plan after others already in progress. In Fort Collins, Colorado, for example, they have set a goal of diverting 50 percent of their waste from landfills. As part of this effort, they have banned the disposal of old computers, TVs, cell phones, and other electronic items, requiring that they be recycled instead. They also make residents pay an additional fee to have extra bags of garbage picked up.

According to Susie Gordon, senior environmental planner for Fort Collins, "The most remarkable difference is that you used to see people with

(continues)

(continued)

15 bags of lawn clippings on the street; you don't see that anymore. The city estimates that almost 30 percent of its waste is now being recycled."

Benita Duran, a CH2M Hill executive who cochaired the group that developed the plan, stated, "This is not a trend or a city competition. It's a global matter of serious concern. The era of denial is over."

There are other major cities jumping on the bandwagon to do their environmental share as well, such as Seattle, Portland, and Chicago. As of February 2009, 911 mayors have signed the U.S. Mayors Climate Protection Agreement, pledging to take action to reduce greenhouse gas emissions by 2012.

John Healy of Seattle's Office of Sustainability and Environment says, "There was an absence of federal leadership on the issue. Now the era of denial is over and we're entering the era of action."

Seattle's goal is a 7 percent reduction in global warming pollution by 2012. At this point, they are not imposing penalties on heavy users of fossil fuels. They are encouraging residents to use biodiesel fuel made from soybeans in their furnaces. Biodiesel is also currently being used in buses and at the airport.

In Portland, Oregon, per capita emissions have fallen 12.5 percent since 1993. This was the year it became the first American city to adopt a goal of reducing greenhouse gases.

higher than the globally averaged temperature increase because of their northern latitude. If CO_2 concentrations double, temperatures there are expected to increase by 5–13°F (3–8°C). Warming temperatures will lead to several ecological consequences. Ecological models predict that warmer temperatures would move the habitat for the northern hardwood forest species northward by at least 100–300 miles (161–483 km) by the end of the next century. The hardwoods will also move higher up the mountains. These are the world famous trees with beautiful, vibrant leaves, whose changing colors attract a huge tourist industry each fall. If these trees migrate, the negative economic impact will be huge. Also soils at higher elevations are often thin and acidic and may not be able to support a forest habitat, in which case the hardwood forest species could completely perish.

Interestingly, not all models project a negative scenario. One particular model predicted that warmer temperatures, combined with increased CO_2 concentrations in the atmosphere, could cause the forests to actually become more productive. In this case, the trees would accumulate and store more carbon as biomass. That, along with an extended growing season and warmer temperatures, could cause an increase in wood production. This model also predicted that the trees would use water more efficiently.

One theory explaining this unexpected growth increase in the northern forest model is that forests are normally limited by nitrogen availability. In the warming scenario, more nitrogen is supplied through the decay of deadwood on the forest floor (which releases nitrogen). The catch to this scenario, however, is that as atmospheric pollution also continues to deposit nitrogen, the capacity of the forests will eventually become saturated. Once this happens, the period of enhanced growth will come to an end, making this benefit short lived.

The models run on the White Mountains indicate vulnerability in several areas. Incidences of extreme or unusual weather, such as periods of winter thaw, intense cold, spring and summer drought, and summer heat waves, can all have devastating impacts. Occurrences of rapid warming have caused significant forest dieback in the past. In the 1980s, Québec had a major dieback of their sugar maples due to a combination of winters with loss of snow cover, followed by deep freezes, followed by summer drought.

Climate change will have a negative effect on three of the region's principal industries—the tourism and recreation industry, the maple syrup industry, and the timber industry. Tourism and recreation are critical to the state economies of the region. According to a study conducted by the Institute for New Hampshire Studies and the Environmental Defense Fund, tourism and recreation represent an annual income of more than $4 billion in New Hampshire, 95 percent of the gross state product. This area attracts large populations of visitors every year eager to view the breathtaking fall foliage displays, to enjoy skiing at resorts, to participate in recreational fishing activities, and to hike the mountain trails and participate in nature walking.

Fall foliage displays will change for the worse as temperatures rise. Some climate models predict that warmer temperatures will continue longer into the fall than they currently do, meaning the leaves may change color in November instead of October. But the changing leaves are governed by both temperature and day length, and different species of trees are affected differently. This could cause an uncoordinated display of fall colors. It may also cause the fall colors to appear less vibrant, hurting their popularity.

The ski industry also plays a significant role in the region's economy. Global warming in the region could cause the ski season to begin later and end earlier. The current ski season lasts from December through April. The larger resorts that have snowmaking capabilities are generally open from two weeks before Thanksgiving to late April. Figuring in the effects of global warming, however, two climate models were tested: one with temperatures rising 3.3°F (2°C) and the other 5.7°F (4°C). The ski season length for these models was predicted to be 16–35 days shorter than normal—a loss of about 10–20 percent of the total season length and income.

Climate change also affects fish habitat, especially in cold-water rivers and streams due to the impact of warmer temperatures and changes in precipitation on stream temperature and flow rates. Cold-water fish species common to the White Mountains region may be unable to survive a significant warming of water temperature. Brook trout, for instance, have strict temperature requirements and would be especially vulnerable to climate change. Stream quality could also be negatively affected by changes in the timing or amount of precipitation. Brown and brook trout have both been identified as vulnerable to extinction if the climate warms.

According to the Environmental Protection Agency (EPA), loss of habitat for cold-water fish may be significant. Rainbow trout, brook trout, and brown trout are the most important cold-water fish for recreational fishing in the New England area, and several states could potentially lose all habitat suitable for these species.

Climate change would also be detrimental to hiking and camping. Warmer temperatures will probably lengthen the hiking season by increasing the number of snow-free days in the mountains, but the quality of the natural experience may be reduced based on higher temperatures, lack of precipitation, dead and dying trees, pest and pathogen outbreaks, more frequent fires, and more frequent droughts. All of these issues may make the White Mountains less aesthetically appealing and less attractive for overall recreational value.

Sugaring, the harvesting of sugar maple sap to produce maple syrup, is a tradition in the White Mountains that dates back to precolonial days. Climate change is having an effect on the maple syrup industry in the region—even in New England as a whole—shortening the season and decreasing profits.

The reason why maple syrup production is at such great risk is that strong sap flow only occurs when there is a sharp difference between day and night temperatures: Optimal sap flow in sugar maple is dependent on a prolonged early spring period with cold (less than 25°F or -3.9°C) nights and warm (more than 40°F or 4.4°C) days. With global warming, there will be less difference between day and night temperatures—what is referred to as the diurnal range. With a decrease in the diurnal range of temperatures, the length (and profitability) of the maple syrup season is threatened. If warming is more pronounced at night, the number of days of optimal syrup flow will decrease. A typical season is about 33 days with 18 of them optimal flow days. Under the effects of global warming, the number of optimal flow days could be reduced to as low as 11, causing a serious economic impact.

In addition, if increased temperatures cause the buds on the branches to break earlier in the season, the sap collected produces syrup that is bitter and unmarketable. Another factor that may harm the maple syrup industry is the condition of the soil and drought brought on by global warming. Sugar maples need moist, rich soils in order to be productive. If increasing temperatures cause the soils to become drier, the health of the trees will become threatened.

The timber industry is also an important component of the region's economy. Timber is harvested for use in furniture, building materials, pulp for paper production, and fuel wood. Under changing climate conditions, many tree species may not be able to reproduce and grow in their current locations. Increased disturbances from fire, pest and pathogen outbreaks, and storm damage could also lead to changes in forest species and type. In addition, as species shift under the influence of changing temperature distribution, there will be many trees that die off, which would decrease their short-term productivity.

Based on climate models, white pine and red oak—the two most important species today in New England—could increase production in a warmer climate. Quality northern hardwoods (sugar maple, ash, and yellow birch) could decline, especially in extreme or unusual weather. Spruce and fir could decline as warming temperatures push the trees upward in elevation. Beech and red maple, which are more tolerant of dry conditions, could increase, but are less valuable as timber resources.

Because of the uncertainties in the models, however, the economic impacts are very difficult to determine. At this point, land managers can use these results as guides. As economic models become more accurate in the future, it will be possible to predict risks and benefits more reliably.

As global warming continues, each geographic location—like the White Mountains—will have cause-and-effect scenarios related to its natural resources, habitats, industry, and quality of life. These are the issues that must be dealt with as temperatures climb and drastically change the environment, another reason why preventing global warming is important.

Impacts to Marine Ecosystems

The Earth's marine environments are also susceptible to the effects of global warming. As evidenced in chapter 5 on the effects of global warming to the marine wildlife and ecosystems in the polar regions, other marine environments are also at risk as temperatures continue to climb. This chapter will focus on the temperate marine and coastal environments found at the Earth's midlatitudes, as well as the marine ecosystems located in the tropical regions, and why rising temperatures are putting some of the Earth's most fragile ecosystems at risk. It will then focus on freshwater environments and the effects global warming is having on wetlands, streams, lakes, and other freshwater environments and the wildlife that lives within them.

TEMPERATE MARINE ENVIRONMENTS

There are several components of the temperate marine environment that will be targeted by global warming, including changes in temperature, shoreline ecology, and major storm tracks. Two elements that are the least complicated to predict are temperature and sea-level rise; others are much more complex depending on scale and degree of interaction between other components in the environment.

It is difficult to use existing climate models to address these issues because they produce outputs that generally depict broader geographical regions than the scales where local storms occur. Models are not refined enough at this point for their spatial resolution to be able to model areas as small as a specific bay or coast.

According to the Intergovernmental Panel on Climate Change (IPCC), by the year 2100, the Earth's near surface temperature averaged worldwide will increase by 2.5–10.4°F (1.4–5.8°C) from 1990 levels. This means that sea surface temperatures will also rise. The area expected to warm the most is the high (polar) latitudes in the winter. The IPCC believes that this will be the highest temperature rise seen in the past 10,000 years. The aspect of the temperature rise that will cause the most significant ecological change in *estuary* and marine ecosystems is the rapid speed with which it is expected to happen, leaving species little time to adapt.

Based on data from the IPCC, globally averaged sea level rose between four and eight inches (10–20 cm) during the 20th century. The IPCC predicts the oceans will rise another 4–35 inches (9–88 cm) between now and 2100. The rise will be the result of both (1) thermal expansion of the current ocean water (the warmer the temperature, the more water expands) and (2) melting from land-based glaciers and ice sheets.

In addition, the IPCC predicts that globally averaged precipitation will increase in the future. Winter precipitation will increase, as well as storms over mid to high latitudes over the Northern Hemisphere. In the temperate marine environments, there will be specific effects from global warming on both the coastal locations and the open ocean areas.

The Effects of Global Warming on Coastal Locations

The Pew Center on Global Climate Change is a nonprofit organization that brings together business leaders, policy makers, scientists, and other experts worldwide to create a new approach to managing the problem of global warming—what they refer to as an extremely "controversial issue." Not slanted in any particular way politically or economically, they "approach the issue objectively and base their research and conclusions on sound science, straight talk, and a belief that experts worldwide can work together to protect the climate while sustaining economic growth."

The Pew Center was established in 1998 and has "issued more than 100 reports from top-tier researchers on key climate topics such as economic and environmental impacts and practical domestic and international policy solutions." The center's leaders and staff hold briefings with members of Congress and administration officials in the U.S. government, as well as with leaders of international governments. They also work with 43 major corporations in business roles to encourage them to help promote practical solutions to solving the global warming crisis. In January 2007, the Center was one of the inaugural members of the U.S. Climate Action Partnership—an alliance of major businesses and environmental groups that calls on the federal government to enact legislation requiring significant reductions of greenhouse gas emissions.

According to the Pew Center on Global Climate Change, in terms of global warming, the biggest impact on estuarine and marine systems will be temperature change, sea-level rise, the availability of water from precipitation and runoff, wind patterns, and storminess. In these often-fragile systems, temperature has a direct and serious effect. For the sea life living within the ocean, temperature directly affects an organism's biology, such as birth, reproduction, growth, behavior, and death. The remainder of this section will further explore the specific effects global warming will have on temperature, sea-level rise, wind circulation, aquaculture, algae blooms and disease, and the long-term effects of coastal building.

Temperature

Temperature extremes (both high and low) can be deadly to living organisms. Many species are vulnerable to temperatures just a few degrees higher than what they are accustomed to. Even increases in temperature as little as 1.7°F (1°C) can seriously harm certain species. For example, from 1976–77, the species of reef fish off Los Angeles diminished by 15–25 percent when temperatures abruptly warmed 1.7°F (1°C). Even if the temperature is not high enough to kill, it can be high enough to negatively influence the organism's life functions such as metabolism, growth, behavior, and physiological factors such as the timing of reproduction rates of egg and larval development.

Temperature also influences where organisms can survive and controls how large a given population can become. One area where this has long been documented is along the West Coast of North America. The nearshore sea surface is predominantly warm and the offshore area is predominantly cool, creating a region that has long been famous as a rich, diverse fishery.

Temperature differences can also influence interaction between species, such as predator-prey, parasite-host, and other relationships that may develop over the struggle for limited resources. If temperatures change the distribution patterns of organisms, it could also change the balance of predators, prey, parasites, and competitors in an ecosystem, completely readjusting balances, food chains, behaviors, and the equilibrium of the ecosystem.

Global warming can also change the way that species interact by changing the timing of physiological events. One of the key changes that it could alter is the timing of reproduction for many species. Rising temperature could interfere with the timing of birth being correlated with food availability of that species. This can be a problem, for example, for bird species that migrate and depend on a specific food source to be available when they reach their breeding grounds. If warmer temperatures have changed the timing and it is a few weeks off of when the food will be available and no longer synchronized with migrating birds, it could leave the birds without available food, threatening their survival.

Temperature also plays an important role with oxygen because it directly influences the amount that water can hold. The warmer the water, the less oxygen it holds. As global warming continues to force temperatures to rise, less oxygen will be available for marine species, which could then threaten their survival.

A significant amount of data has already been collected on these variables. According to the Pew Center, areas that still need much more data collected and research applied concern the ocean temperatures influence on the interactions among organisms, such as predator-prey relationships, parasite-host relationships, and the interplay of species concerning competition for resources.

It is also possible to model the effects of sea-level rise in shallow continental margins, such as flooding wetlands and shoreline erosion. What is more difficult to predict and needs more research are the global warming effects on precipitation amounts, wind patterns, and intense weather. Precipitation directly affects estuaries because it affects the runoff into estuaries, which influences estuarine circulation.

Sea-Level Rise

Sea-level rise will mount from the melting of glacial and polar land ice. According to the Pew Center, the effects of sea-level rise will vary by location, how fast the sea level rises, and the biogeochemical responses of the individual ecosystems involved. One of the areas identified as being the most susceptible to damage from sea-level rise is the low-lying, flat wetlands located in the middle and south Atlantic and the Gulf of Mexico (involving states such as Alabama, Florida, Mississippi, Georgia, and Texas). A wetland is an area on shore that has a wet, spongy soil. These areas are also referred to as swamps, marshes, and bogs.

As sea levels rise, ocean water will submerge and erode the shorelines. In the natural areas covered with marshes and mangroves (common to Florida and the Gulf states), sea levels will flood the wetlands and waterlog the soils. Because the plants that live in the wetlands, which do not contain salty water, are not accustomed to salt, the salty ocean water will kill them. Because wetlands provide habitat for wildlife, including several migrating birds, this would also destroy their habitat.

Based on research by the Pew Center, if the wetlands located along the Gulf of Mexico are not bordered by human development (such as homes) and the inland slope is relatively flat, and the sea-level rise is gradual, they would have a chance of surviving by migrating further inland to escape the inundation of salt water. The Pew Center states that "these areas would be safe at least at the rates of rise currently projected for the next 50 to 100 years."

The areas that would be hard hit are those that cannot migrate inland because urbanization has grown right up to their shoreline, effectively removing any potential wetland habitat. This is very detrimental to the environment because wetlands are an important part of the biological productivity of coastal systems. Marshes provide many critical services; they function not only as habitats for wildlife, but as nurseries for breeding and raising young and as refuges from predators. Wetlands function as part of an integrated system. If they are jeopardized, their loss will affect the availability and transfer of nutrients, the flow of energy, and the availability of natural habitat needed by multitudes of organisms already living there. One of the most unfortunate losses will be those areas where rare, threatened, or endangered plant and animal species live, such as the American alligator, Florida black bear, West Indian manatee, Florida panther, southern bald eagle, snowy egret, and roseate spoonbill. If invading salt water destroys these habitats, all these species could become extinct.

Wind Circulation

Another consideration in the marine environment is wind circulation. Winds are created by the uneven heating at the Earth's surface, specifically the high thermal gradients between the equator and the poles. Under global warming conditions, however, the polar regions will be subjected to higher temperatures, thereby reducing the temperature gradient between the equator and the poles. This could cause a weakening in the overall wind circulation around the Earth.

The winds drive the surface currents of the Earth's oceans. A slow-down of the ocean surface circulation could negatively affect the structure and function of both the open ocean and the near shore ecosystems. Weak currents would make it difficult for any currents to transport nutrients where they need to be deposited.

If wind speed and direction are altered, which is what experts at the Pew Center have predicted will occur with global warming, the productivity of estuarine and marine systems will also be affected. One of the reasons why some coastal areas are so rich and productive in fishery resources, such as the West Coast of the United States, is because of the coldwater *upwelling* that occurs along the coastal margins. The thermal properties and circulation patterns function to bring the cold water from the deep ocean waters to the surface along the coastlines, supplying the coastal fisheries with an abundance of fish. Nutrients that are delivered from these deeper waters to the surface help provide for the phytoplankton along the coasts. The upwelling process is controlled by the winds that blow along shore. Upwelling can be minimized, however, if the water column is stratified by differences in temperature or salinity, which act as a barrier to upward movement through the water column.

The effects of this type of scenario were played out off the coast of California from 1951 to 1993. The sea surface warmed during this period 2.5°F (1.5°C), resulting in a decreased upwelling of nutrient-rich cold waters. This caused a 70 percent decline in the abundance of zooplankton, which in turn harmed the coastal food webs, negatively affecting fish and birds.

There are different opinions on this matter, however. According to Andrew Bakum of the National Marine Fisheries Service of the National Oceanic and Atmospheric Administration (NOAA), his research leads him to believe that global warming may actually increase coastal upwelling in some areas. He believes this is possible because most greenhouse gas models project more warming over land than over open oceans, which would increase the temperature contrast between the land and the ocean. This resultant contrast would strengthen the low-pressure cells that usually form over land that are adjacent to offshore high-pressure cells. The relationship of forces between these two pressure cells would create alongshore winds that would create a strong upwelling.

On the East Coast of the United States, circulation patterns could slow down transport of nutrients and fish species, such as blue crabs and bluefish, which would lower the abundances of these particular species along the coasts and within estuaries. When their populations are decreased, the coastal ecosystem becomes threatened.

Aquaculture

Coastal aquaculture, the farming of fish from the ocean in special facilities, has become a key food source in recent years. According to the Food and Agriculture Organization (FAO), global aquaculture production has been increasing since the 1950s at a growth rate of about 10 percent per year since 1990. By 2030, the FAO projects that aquaculture harvests will be greater than capture harvests. Marine aquaculture is a steadily growing industry.

The effect that global warming will have on aquaculture is mixed. On one hand, the higher temperatures could enhance growth rates of aquaculture species and could make it possible to run aquaculture operations in locations that are currently too cold. On the other hand, areas that are suitable now for aquaculture may become too warm for existing species to survive. The Pew Center's research suggests that as conditions change, if the temperatures rise slowly enough, aquaculture operations should be able to keep up with the changes and keep negative effects to a minimum.

Algae Blooms and Disease

One major drawback of aquaculture facilities is that they affect local ecosystems when their concentrated wastes pollute the surrounding environment, encouraging algae blooms. If global warming becomes an issue, algae blooms will become more common. Warm water holds less oxygen and accelerates the microbial decomposition of the aquaculture wastes. Oxygen concentrations are further lowered, exacerbating the problem. As global warming continues and temperatures climb upward, disease will become more prevalent, distributing the pathogens to more areas. Not only does this affect ocean life, it also affects humans.

An example of this has been illustrated with oysters. The protozoan *Perkinsus marinus* (Dermo) is a pathogen that threatens the health of oysters. Cool ocean temperatures keep infection by this pathogen to a minimum. As water temperatures have risen in recent years, the spread of this pathogen has been documented spreading to the northeastern states of the United States. Based on this evidence, Dr. T. Cook of Rutgers University's Institute of Marine and Coastal Sciences says long-term climate changes may produce shifts in salinity and temperature that enable pathogens to spread. Climate change could also affect the distribution of other aquatic diseases in a similar manner.

Warmer coastal waters along with eutrophication (an increase in chemical nutrients—usually compounds containing nitrogen or phosphorus—in an ecosystem) can also increase the intensity of harmful algal blooms, which can destroy habitat and shellfish nurseries and are also toxic to both marine species and humans. Unfortunately, these blooms have been recently increasing worldwide, possibly because they are correlated to global warming. One of the concerns is that consumption of shellfish that have ingested harmful algae can cause neurotoxic poisoning in humans.

According to the Pew Center, there are two schools of thought concerning species adaptation and survival in marine ecosystems. One group thinks that marine ecosystems will have far fewer extinctions than terrestrial ecosystems because marine species have wider geographic temperature ranges and the ability and opportunity to migrate to new habitats. The other group thinks that the lack of evidence of recent marine extinctions is simply because the ocean systems are so vast that scientists are not aware of them, and there is not enough data collected about the oceans.

Long-term Effects of Coastal Building

According to an article in the *New York Times* on July 25, 2006, entitled "Climate Experts Warn of More Coastal Building," 10 climate experts from around the United States say that the unchecked coastal development in geographic locations that are vulnerable to hurricanes and other forms of severe weather, along with the lack of government regulation on development, is the biggest reason why extreme, violent weather is currently causing so much human suffering, loss of life, and property damage. The article states that the opinion of the experts is that: "Whatever the relationship between hurricanes and climate, hurricanes are hitting the coasts and houses should not be built in their path."

One of the main problems is social/political in nature. Because coastal areas are popular places to live, pressures are put on states and Congress to obtain discounted insurance for property known to be in harm's way. The climate experts claim that reimbursement for property loss wrongly encourages victims of disastrous weather to build again in the wrong place. "Federal disaster policies, while providing obvious humanitarian benefits, also serve to promote risky behavior in the long run." The climate scientists also stressed that storms like Katrina "are inevitable even in a stable climate."

One of the scientists, Philip J. Klotzback, a hurricane researcher at Colorado State University who does not support the notion that global warming is spurring stronger or more frequent hurricanes, stated, "The social and economic trends are completely clear. There is likely to be destructiveness in *tropical storms* anyway because more people now live in vulnerable coastal areas."

Kerry A. Emanuel, a climatologist at the Massachusetts Institute of Technology who does believe that the building energy of hurricanes in recent decades is related to human-driven warming of the seas, also warns people from building in these areas.

The Effects of Global Warming on Open Oceans

Because the oceans are so vast, much of the predictions made about the Earth's open oceans in light of global warming have been generated by computer models. Because the two most important functions that govern the behavior of the ocean are temperature and circulation, these calculations are fairly straightforward in computer models and readily calculable. Another thing that makes it possible to model the oceans using computers is that humans have not had as large an impact on the oceans as they have on land.

The Thermohaline Circulation, the Great Conveyor Belt

Because of the interactions of temperature and salinity in the world's oceans, movement is generated that links them all. This linking of the oceans allows heat energy to be transported globally. Cold water is much denser than warm water, so it sinks to the bottom of the ocean while the warmer water rides above it. Water that is high in salt content is also denser, which forces it below water that is lower in salinity. These differences in density are what generate the movement of the oceans.

The most important movement is the thermohaline (*thermo* for "temperature" and *haline* for "salt content") system, commonly known as the great ocean conveyor belt. This system moves cold dense seawater from the North Atlantic Ocean surface to deep water and then continues through the Indian and Pacific Oceans and back to the Atlantic.

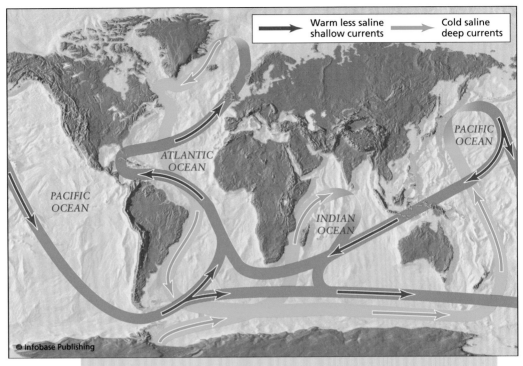

The thermohaline circulation, or great conveyor belt, distributes water of varying temperatures and salinity throughout the Earth's oceans, transferring nutrients and heat energy.

This process takes centuries for an entire cycle to be completed, and it is one of the most important currents in the ocean for regulating heat on the Earth's continents. This is the current that picks up heat in the equatorial region, rises to the ocean's surface, and travels northward via the Gulf Stream/North Atlantic Current past the west coast of Europe, moderating the climate there, making it much milder than it otherwise would be at its northern latitude.

As the current reaches its northern limit in the Arctic, it cools and sinks and takes with it the nutrients as well as oxygen and carbon dioxide (CO_2) that it absorbed while it was at the surface of the ocean, creating a sink for CO_2. Deep-sea organisms are able to use the nutrients and oxygen, but the CO_2 does not get used by plants in photosynthesis because the ocean depths are too dark.

If global warming intensifies and freshwater from melting glaciers and ice caps is added to the Arctic waters, it will dilute the salinity and slow or stop the vertical mixing between the ocean surface and the deep sea. The result of this would be to slow, or even halt, the conveyor belt. If this were to happen, warmth would no longer be delivered to Europe via the Gulf Stream, which could put Europe into an ice age.

There would be many other effects as well. Diminished vertical mixing between the ocean's surface and depths would reduce the upwelling in temperate and subtropical latitudes. Oxygen would not be effectively transported from the surface to the ocean depths. Over time—centuries—the deep ocean waters would become devoid of oxygen, and they would either become hypoxic (low in oxygen concentrations) or anoxic (completely lacking in oxygen).

The dispersal of nutrients from the surface to the depths would also be impaired, leaving deep-sea animals, including commercial fish and squid, without nutrition. This could destroy the fishing industry.

As far as the present rate of CO_2 uptake by the oceans, the Pew Climate Center believes there is disagreement as to whether CO_2 would be less likely to be absorbed if the ocean depths were not mixing effectively, removing the ocean as a CO_2 sink. This would increase CO_2 concentrations in the atmosphere or it would cause an increase in biological productivity in the upper ocean layers and increase the CO_2 uptake.

According to the IPCC in their third report, general circulation models (GCMs) predict a wide range of global ocean circulation responses to global warming, ranging from "no response" to a "40 percent decline in circulation before 2100." According to Dr. Thomas F. Stocker of the University of Bern's Physics Institute, "changes in the thermohaline circulation are likely with global warming, but the extent of the changes are uncertain at this point."

Dr. Peter Robert Gent, chairman of the science steering committee for the Community Climate System Model, working with the National Science Foundation and the U.S. Department of Energy, says he "does not expect a collapse of the thermohaline circulation during the 21st century because of the feedbacks that help stabilize circulation in response to warming." He warns, however, that "presently, little is known about the feedback mechanisms that affect the thermohaline circulation pattern." Another important consideration, he says, is "that these projections are made under the assumption that atmospheric CO_2 levels do not exceed a doubling of preindustrial levels before 2100. Models that were run basing a CO_2 level above double the preindustrial levels before 2100 did simulate the collapse of the thermohaline circulation by 2100—which could put Europe in an ice age." These are still major unknowns concerning global warming and much more research needs to be done.

Ocean Acidification

Ocean acidification is the ongoing decrease in the pH value of the Earth's oceans caused by their steady uptake of *anthropogenic* (human-caused) CO_2 from the atmosphere. According to the German Advisory Council on Global Change, the oceans presently store about 50 times more CO_2 than the atmosphere and 20 times more than the terrestrial biosphere and soils. The ocean is not only an important CO_2 reservoir, it is also the most important long-term CO_2 sink.

Due to pressure differences between the atmosphere and seawater, some of the anthropogenic CO_2 dissolves in the surface layer of the ocean. Over periods of time—decades to centuries to millennia—the CO_2 is carried into the deep sea by ocean currents. According to Christopher L. Sabine of NOAA's Pacific Marine Environmental Laboratory, the ocean is presently taking up 2 Gt (gigatons) of carbon annually, which according to the IPCC is the equivalent of roughly 30 percent of the anthropogenic CO_2 emissions. (A gigaton is one billion tons [900,000,000 metric tons].) The IPCC calculates that between 1800 and 1995, the oceans absorbed about 118 Gt of carbon. This corresponds to about 48 percent of the cumulative CO_2 emissions from fossil fuels, or 27–34 percent of the total anthropogenic CO_2 emissions, including those from land-use changes such as deforestation.

Currently, anthropogenic CO_2 can be tracked to a depth of 3,281 feet (1,000 m) in the ocean. It has not sunk to the bottom of the ocean yet because the vertical mixing process is so slow that it takes a long time. The North Atlantic region is different, however. Because the vertical mixing happens much more readily with the currents there, anthropogenic CO_2 has been traced to a depth of 9,843 feet (3,000 m).

In the ocean, CO_2 behaves differently than it does in the atmosphere. In the atmosphere, it is chemically neutral; in the ocean it is chemically active. Dissolved CO_2 contributes to the reduction of the pH value and causes an acidification of seawater, which is a property that can be easily measured.

Since the onset of the Industrial Revolution in the 1700s, the pH value of the oceans has dropped (become more acidic) by about 0.11 units. Starting from a slightly alkaline, preindustrial pH value of 8.18, the acidity of the ocean has increased at the surface. Based on modeling done by the IPCC, if atmospheric CO_2 concentrations reach 650 parts per million (ppm) by the year 2100, a decrease in the average pH value by 0.30 units can be expected compared to preindustrial levels. If concentrations increase to 970 ppm, the pH level would drop by 0.46 units.

The scale of changes will vary regionally, which will affect the magnitude of any biological effects. Based on the following data from the Royal Society in London, they believe that ocean acidification

> . . . is essentially irreversible during our lifetimes. It will take tens of thousands of years for ocean chemistry to return to a condition similar to that occurring at preindustrial times (about 200 years ago). Our ability to reduce ocean acidification through artificial methods such as through the addition of chemicals is unproven. These techniques will at best be effective only at a very local scale, and could also cause damage to the marine environment. Reducing CO_2 emissions to the atmosphere appears to be the only practical way to minimize the risk of large-scale and long-term changes to the oceans.

All the evidence collected to date points to ocean acidification being caused by human activity—burning fossil fuels, deforestation, and land-use change. The magnitude of acidification can be predicted accurately. What cannot be predicted yet is the physiological effects on various organisms. Headway has been made, however, on discovering the ramifications of acidification on the process of calcification—the process by which organisms such as corals and mollusks make shells and plates from calcium carbonate.

The tropical and subtropical corals are expected to be the worst affected. This could destroy coral reef habitats. Phytoplankton and zooplankton, which are major food sources for fish and other animals, will also be negatively affected. This is one area of global warming where research is still in its infancy and needs much more to be done in order to understand both the short- and long-term effects.

Seaweed—A Carbon Sink?

Recent studies indicate that seaweed may be capable of sucking CO_2 out of the atmosphere at rates comparable to the world's rain forests. According to Chung Ik-Kyo, a South Korean environmental scientist, "The ocean's role is neglected because we can't see the vegetation. But under the sea, there is a lot of seaweed and sea grass that can take up carbon dioxide."

This concept has such potential, in fact, that 12 Asian-Pacific countries are currently working together to calculate how much CO_2 is being absorbed from the atmosphere by plants and to increase carbon storage through carbon sinks.

In 2007, the United Nations Climate Change Conference in Bali brought together 10,000 participants from 180 countries and adopted the Bali Road Map, various decisions toward reaching a secure climate future. A major issue addressed was identifying those things that can be potentially used as carbon sinks to remove carbon from the air.

One possible sink, according to Chung Ik-Kyo, is seaweed. There is a tremendous amount of seaweed produced and available each year—8 million tons (7.3 million metric tons)—harvested from wild or cultivated sources. This theory has met with criticism, however. Some believe it will be a challenge to keep the carbon, once it is absorbed in the seaweed, from reentering the atmosphere at some point. Another negative side effect is the uncertainty about exactly how an increase in seaweed production would affect ocean navigation or fisheries.

The world's current biggest producers of seaweed are China, South Korea, and Japan. Those that favor the plan of using seaweed say that seaweed and algae's rapid rate of photosynthesis (the process of turning CO_2 and sunlight into energy and oxygen) make them the perfect candidates for absorbing carbon.

According to Lee Jae-young of South Korea's fisheries ministry, "Seaweeds can absorb five times more carbon dioxide than plants on land."

John Beardall with Australia's Monash University says, "The oceans account for 50 percent of all the photosynthesis on the Earth. These are very productive ecosystems."

In addition to using seaweed as a CO_2 storage, these same researchers have also suggested that it be used to produce clean-burning biofuels. Skeptics of this concept say that the reason trees are so effective at carbon storage is because they live for many years. Seaweed is grown and harvested in terms of a few months. They feel that this would make this type of carbon storage difficult to measure and control.

Other critics wonder if removing water from the seaweed as it is converted to fuel would require a larger input of energy than that which would be received, making it environmentally unwise. Chung responds by saying, "The idea is still in its infancy. In terms of ball games, we are just in the bullpen, not the main game yet." Currently, South Korea and Japan are the leaders in this type of research.

TROPICAL MARINE ENVIRONMENTS

The world's tropical and subtropical marine environments represent some of the most diverse habitats on Earth. Often characterized by reef-building corals, the complex arrays of marine inhabitants that occupy these waters have developed many strategies for survival. Their ecosystems are so complex that a delicate balance is needed to protect these numerous marine resources, while also accommodating an economy centered on commercial fisheries and recreation.

The tropical marine environment is also subjected to many of the same environmental concerns as the temperate marine environment, such as temperature, sea-level rise, wind circulation, and algae blooms. An additional risk in the tropical habitat is warming's negative impact on reefs and corals. These ecosystems represent some of the most fragile on Earth and are some of the hardest hit by global warming.

Fragile Ecosystems—Reefs and Corals

As global warming causes the Earth's tropical oceans to heat up and become more acidic and increases the incidence of strong storms, the world's coral reefs are taking a beating. Rod Fujita of Environmental

Coral reefs are often referred to as the rain forests of the ocean because they provide habitat for an extremely diverse collection of life-forms. Global warming is threatening their existence. *(NOAA)*

Defense Fund (EDF) says, "Coral reefs may prove to be the first ecological victims of unchecked global warming."

According to the U.S. Commission on Ocean Policy, losing coral reefs would translate into huge economic losses in coastal regions. They currently provide about $375 billion each year in food and tourist income.

Destruction of reef systems also represents an ecological disaster. Coral reefs are sometimes called the rain forests of the ocean because they provide habitat for a rich diversity of marine life, including reef fish, turtles, sharks, lobsters, anemones, sponges, shrimp, sea stars, sea horses, and eels.

Coral reefs attract scuba divers from around the world each year to swim among the beautiful, otherworldly shapes and color combinations. Corals actually obtain their food and color from tiny algae called

As global warming heats the water, it endangers the wildlife and coral that live in the tropical regions, such as this sea turtle. *(NOAA)*

zooxanthellae that live in them. Corals have a very narrow temperature tolerance range, meaning they only thrive within a narrow range of heat and cold. In fact, if the water increases only 1.8°F (1.1°C) above the typical maximum summer temperature, it can cause coral to expel their algae and turn white, through a process called bleaching. If bleaching continues for a prolonged period of time, the corals will die.

Sadly, there have been several highly destructive or fatal bleachings to the world's coral. In 1997–98, during one of the warmest 12-month periods ever recorded, there was massive documented bleaching of the world's corals. Sixteen percent of the Earth's reefs suffered severe damage. The warming caused ancient, thousand-year-old corals to die off. Doug Rader, a scientist with EDF, says, "Within a century, very large portions of coral reefs could be gone. It seems hard to believe that it is happening—and happening on our watch."

Rod Fujita chooses to hope for a brighter side: "Corals are sensitive, but also very resilient—if conditions are right. If we can reduce some of the other direct stresses from human activities on coral reefs, like pollution from diffuse sources, it may enable reefs to cope better with threats like climate change. Creating more protected areas for coral reefs may help them better withstand the rigors of too-warm water and be less vulnerable to extinction, as well. But even so, the number of corals that can adapt to or withstand such dramatic, rapid changes may be just a tiny fraction."

In an article in the *New York Times,* it was confirmed that increasing acidity in the oceans hurts the reef algae as well as the corals. Ilsa B. Kuffner of the United States Geological Survey (USGS) was involved in experimentation illustrating that when seawater absorbs CO_2 it leads to lower saturation levels of carbonate ions, which reduces calcification. Calcification is the process that corals use to make their hard skeletons. Ocean acidification also has a negative impact on crustose coralline

Coral reefs are popular for scuba divers for both recreational diving and scientific research. As global temperatures rise, these reefs will become destroyed. *(NOAA)*

algae, which is another important reef builder that acts as a cementlike substance that bonds the reefs and helps the reef ecology.

The result of Kuffner's experiment proved that the specimens subjected to higher heat and acidity became soft algae, not healthy, resistant algae.

Global Warming Stress to Coral Reefs

The World Wildlife Fund (WWF) reports that coral reefs around the world have been severely damaged by unusually warm ocean waters. They predict that less than 5 percent of Australia's Great Barrier Reef will remain by 2050 if the world fails to reduce CO_2 emissions. They project that "if the present rate of destruction continues, a good proportion of the world's coral reefs could be killed within our lifetime."

WWF has identified both the Seychelles Islands and American Samoa as locations under high stress for coral bleaching. The United Nations Educational, Scientific and Cultural Organization (UNESCO) recognizes the Seychelles Islands as a natural World Heritage Site. They have a high diversity of coral and support rare land species, such as the giant tortoise. In addition to increasing ocean temperatures, these areas are also threatened by global warming because of more frequent tropical storms (which could break up the coral) and more frequent rains, flooding, and river runoff (which deposits sediments in the ocean).

According to the National Aeronautics and Space Administration (NASA), reef habitats are so complex and warrant so much exploration, "that for marine biologists, the destruction of the reefs has proven to be as frustrating as it is heart breaking. At the rate the reefs are disappearing, they may be beyond repair by the time a comprehensive plan to save reefs can be put into place."

Scientists at NASA's Goddard Space Flight Center and at several universities around the world have at least a partial solution. They have been interpreting satellite imagery of the world's oceans obtained from *Landsat 7* and other high-resolution sensors to map the shallow waters around the ocean's margins, creating maps of the reef environments. With funding, they expect to soon have a comprehensive map of the world's reef systems that they can use to identify large-scale threats to the reefs.

In the meantime, Abdul Azeez Abdul Hakeem, a Maldivian scientist, is dedicated to helping corals on Maldives—a nation of 1,200 islands in the Indian Ocean—survive global warming. Azeez is the director of conservation for an eco-friendly resort called Banyan Tree Maldives. Besides being a premier vacation resort, it also hosts a world-class marine laboratory managed by Azeez. The staff spends most of their time studying the coral and maintaining the ecosystem—the reefs provide homes for many exotic fish and attract travelers from all over the world.

Azeez became involved with coral in 1998 when a strong El Niño warmed the ocean and put the coral at risk. "It rose to about 91°F (33°C), so 33 was boiling hell, and about 80 to 90 percent of the corals in the Maldives died. I never believed that an entire region could be wiped out. No one believed that this could happen until it hit us. Then only I also realized yes we are in danger because of global warming and this can happen again and again."

When that happened, Azeez began looking for ways to protect the coral. He also knew that corals on an artificial reef nearby had survived the 1998 El Niño. That reef was an experimental design that used electricity. At that time, no one knew how it had protected the reef from the extreme heat.

Azeez took the design idea and built an electric reef from steel bars and wired it to a power source on the beach at the Banyan Tree Resort. The small electric current caused the minerals from ocean water to build up on the steel, forming a thick limestone crust that is perfect for coral. The electric reef was situated on the far side of the island, submerged in 15 feet (5 m) of water.

The reef is maintained as a type of underwater topiary. An immense variety of corals grow on it like an underwater garden; all thriving in their sheltered habitat.

"Finger corals, hard corals, massive corals. We have tried to plant as many species as we can," Azeez explains. "The reef also attracts both predators and prey, just as a healthy reef ecosystem does."

Azeez considers his electric reef as a type of "greenhouse for corals." He believes it will keep a critical mass alive when the next El Niño strikes. Azeez claims, "You can take pieces from the corals on this structure to that one and make your own garden again." This transplant process seems to be working for Azeez in the Maldives. "Our enemy, the real threat," says Azeez, "is global warming."

In a report in National Geographic News, in 1995, it was stated that roughly 10 percent of the coral colonies in Belize had died. "This coral bleaching is pretty solidly tied to rising ocean temperatures," said Melanie McField, a Belize-based reef scientist with the WWF. "It's a fact that global temperatures have risen. There's lots of data and little argument that increased ocean temperatures are the primary agent of bleaching. As for tying overall temperature increases to overall global warming, there is still some debate, but less every year. I think the majority of scientists agree that global warming is happening and that it's the root cause of these coral bleaching events. Rather than throwing up our hands and saying 'we can't control that,' we've got to be even more diligent and try even harder to control local impacts such as pollution and overfishing," McField said.

Even in view of the unstoppable damage to reefs in the future, conservationists say that given the complex factors affecting coral health, there is still a lot that can be done to help reefs recover if action is taken now.

Protection and Conservation

As the world becomes more aware of the perilous condition of the fragile marine habitat, concern is growing over the future condition and existence of these beautiful areas. As more is learned about what is happening to oceanic ecosystems, more attention is being given to the creation of protected marine areas in order to conserve, manage, and protect ocean resources so that they will not become extinct. The location of conservation areas needs to be carefully chosen to be able to protect and provide for the greatest diversity of species.

On June 16, 2006, former president George W. Bush designated the Northwest Hawaiian Islands a national monument. This created the largest marine protected area in the world, providing important habitat for thousands of marine species that rely on it for survival.

According to Steve McCormick, past president of the Nature Conservancy, "We commend the administration for its foresight and leadership in protecting this incredible area. Designating the northwest Hawaiian Islands as a national monument will ensure that this national treasure will remain healthy and intact for generations to come."

This preservation area includes 5,019 square miles (13,000 km²) of coral reefs, comprising 70 percent of all coral reefs in the United States.

According to National Geographic News, a consortium of Latin American nations, conservation groups, and United Nations agencies are creating one of the world's largest marine protected areas to be known as the Eastern Tropical Pacific Seascape. The new reserve will cover 521 million acres (211 million ha) of ocean from Costa Rica's Cocos Island to Ecuador's Galápagos Islands. The purpose of the preserve is to protect a wide range of ocean species, expand existing marine reserves, and consolidate current and planned conservation efforts. It will be the largest marine conservation area in the Western Hemisphere. The region already has these currently listed World Heritage Sites: Galápagos Islands and Marine Reserve (in Ecuador), Cocos Island (Costa Rica), and Coiba National Park (Panama).

There are several conservation organizations in existence today that are geared toward helping promote healthy ocean ecosystems. Fortunately for the environment, as well as future generations, tens of thousands of people choose to get involved and make a difference.

FRESHWATER ENVIRONMENTS

Only 2.58 percent of the water on Earth is fresh. The largest share of that—1.97 percent—occurs as ice in the world's glaciers and ice caps. The remaining freshwater—0.61 percent—is stored in lakes, rivers, and groundwater. The Earth's climate—the processes of precipitation, evaporation, and water vapor transport—determines the amount and distribution of freshwater at any given time on Earth.

Without freshwater, life could not exist. Freshwater is the most essential resource. Wetlands and river systems provide food and water for life. Wetlands are nature's natural filters of harmful substances and serve to purify water. It is the availability of water that partly determines the distribution of major biome types. Humans depend directly on freshwater for drinking and cooking, irrigation (agriculture), industry, transportation, fisheries, and recreation.

IPCC Assessment

According to the IPCC, while growing populations are already overtaxing freshwater supplies in many parts of the world, global warming will make freshwater supply even scarcer in certain areas. Warming temperatures will also have consequences. Winter ice cover of streams and lakes will decline, and there will be a trend toward later freeze and earlier ice

WHY ESTUARIES ARE IMPORTANT TO THE ENVIRONMENT

Estuaries are extremely productive ecosystems important to near-shore coastal environments. They are semi-enclosed bodies of water found in areas such as lagoons. These are the buffer areas where seawater from the oceans and freshwater from the land meet.

Because estuaries are areas of flux between two different types of interacting bodies of water, their salinity can vary over time. For example, if the land has a high degree of runoff water flowing into the estuary, it makes the water in the estuary lower in salinity. Conversely, in years of drought when there is little freshwater runoff, the estuary is more saline. Because of this variability, the wildlife that uses estuaries must be able to adapt to these temporal changes in salinity.

Drainage from rivers on land can deliver essential nutrients to estuaries, but they can also deliver harmful pollutants. For instance, if agricultural areas with high levels of herbicides are upstream and these chemicals get into the runoff water that is carried to the estuary, they can cause serious pollution damage, possibly killing the life within that ecosystem. Contaminants can include soil, chemicals, sewage, and inorganic debris. Items such as the plastic six-pack soft drink can holders can entangle wildlife such as flamingos, causing tragic, unnecessary deaths.

Vegetation commonly found in estuaries includes grasses, mangroves, and microscopic algae. Marshes and mangroves rely on sediment washed from the land into the estuary to build up a sediment base in which to grow. Over time, the sediments collect, become deeper, and allow the mangrove to grow into a thick barrier. Once they become established, these barriers protect the adjacent land and human populations breakup, as is already happening in Europe. The timing and duration of freeze and the breakup of ice are important because they affect both biological and ecological processes. If precipitation decreases, the water flow rates will drop, and this could affect lakes and streams and lead to changes in habitat and breeding locations of aquatic flora and fauna. The IPCC has also determined that hydrologically isolated systems, such as

against destructive storm surges caused by strong coastal storms and hurricanes.

Marshes also intercept nutrients and contaminants carried in runoff and help protect the health of the nitrogen-sensitive sea grasses that play a critical role in keeping the balance of coastal food webs in order. Nutrient runoff supports high estuarine plant production, which supports high animal production—which includes commercial fisheries. According to Dr. Edward D. Houde from the University of Maryland Center for Environmental Science, Chesapeake Biological Laboratory, and Dr. Edward Rutherford from the Institute for Fisheries Research, Michigan fish production is influenced by freshwater and nutrient inputs in complex ways. They estimate that whereas estuaries represent only 0.5 percent of the world's marine environment, they support about 5 percent of global fish production.

Based on information from the National Safety Council's (NSC) Environmental Health Center, about 75 percent of commercially harvested fish and shellfish, as well as species representing up to 90 percent of recreational catches, depend on estuaries for spawning and feeding, as nursery habitat for their young, or as migration routes to or from spawning or feeding habitats.

Estuaries also help the economy because they are used for recreational activities such a boating, hunting, and bird-watching. Migrating birds depend on them as a food source. The Gulf of Mexico wetlands, for example, accommodate about 75 percent of all the migrating waterfowl in the Central Flyway.

Unfortunately, with increasing population and urbanization occurring along coasts, more than 50 percent of the inland and coastal wetlands in the United States have been destroyed since the 1780s.

wetlands and topographical depressions, would be the most vulnerable areas to global climate change. Areas along larger rivers and lakeshores would feel the least impact.

As freshwater supplies diminish with increasing temperatures and lack of precipitation, competition for diminishing freshwater resources will increase. Even if precipitation increases in some areas during the winter, if it cannot be stored as groundwater and runs off the surface when it melts in the spring, it will be unusable. Summers are predicted to become warmer and drier, which will also lead to a deterioration of freshwater ecosystems.

Impacts to Freshwater and Wetland Ecosystems

Global warming has been determined to be devastating to freshwater aquatic ecosystems based on the results of a study conducted by scientists from Colorado State University (CSU), East Carolina University (ECU), and Louisiana State University (LSU). In their report, titled "Aquatic Ecosystems and Global Climate Change," the researchers believe that global warming will be devastating to trout, salmon, and several species of aquatic plants and animals. CSU researchers forecast significant shifts in fish habitats, decreasing water quality, and disappearing wetlands. Rivers, lakes, and wetlands will all be seriously affected.

N. LeRoy Poff, a freshwater biologist at CSU, reports, "Wetlands and aquatic ecosystems are quite vulnerable to climate change. Projected increases in the Earth's surface temperature due to global warming are expected to significantly disrupt current patterns of aquatic plant and animal distribution and to alter fundamental ecosystem processes, resulting in major ecological changes."

Two of the cold-water fish that are expected to suffer most under increased global warming conditions are trout and salmon. Both species will likely disappear from their current geographic ranges. Species that attempt to migrate either northward or higher may not be able to because of urbanization blocking the way. They may become extinct because there is nowhere else to go. It will be a different scenario for those freshwater fish that thrive in warm-water environments, however. Biologists expect that fish, such as largemouth bass and carp, will actually expand their geographic range throughout the United States and Canada.

Rising water temperatures—even of just a few degrees—will have a significant effect on wildlife in terms of their adaptation requirement. For instance, a 5.7°F (4°C) rise in surface temperature would require aquatic species to migrate 422 miles (680 km) northward to maintain the same thermal habitat conditions that they initially enjoyed before the rise in temperature.

According to CSU, water quality will also become an issue. It will decline because there will be a reduction in springtime melt and runoff,

Flamingos are commonly seen in estuarine environments, such as natural lagoons. As sea level rises, these habitats will be negatively affected, threatening the health of the wildlife that live in these areas. *(Nature's Images)*

as well as higher temperatures. Higher temperatures encourage the growth of algae. The more prolific the algae, the lower the amount of dissolved oxygen there is in the water.

In the study, Poff also stressed that "reduced oxygen in lake waters due to increased production of blue-green algae can lead to the loss of large predatory fish and have negative effects on the food chain." In addition, he says that "climate change could have other subtle effects on aquatic ecosystems as precipitation patterns change. In the coming century, areas accustomed to snow may instead get rain in the winter, causing floods that destroy fish eggs left in streams and leaving little snowpack to sustain rivers during the dry summer months."

The authors of the report—N. LeRoy Poff (freshwater biologist) of CSU, Mark Brinson (wetlands biologist) of ECU, and John Day, Jr. (estuarine biologist) of LSU—agree that the effects of global warming are very difficult to predict because the exact temperature changes at this point are based only on the most accurate climate models developed to date. They also agree that "in the face of inevitable climate change, humans can take actions to minimize the risk of ecosystem disruption."

Ecosystem and Human Adaptation

According to the Pew Center, the ability of aquatic ecosystems to adapt to climate change is very limited. Where animals and plants will need to migrate northward or higher in elevation, aquatic species differ in their dispersal abilities. If they are not able to migrate, they will be doomed. Many aquatic species are spatially isolated due to human activities and interference, making it more difficult for them to adapt. The extinction of many local species is to be expected across all aquatic ecosystems. In addition, it is unknown at this point what will happen when species do try to migrate and come into contact with new, nonnative, more resilient, introduced species for the first time.

The Pew Center stresses that one of the critical uncertainties in projecting future aquatic ecosystem responses to global warming is how humans will interact with these ecosystems during the changes. Human activities have already severely interfered with aquatic ecosystems through the building of dikes, levees, and reservoirs, as well as pumping significant amounts of groundwater. Over the years, these have all modified natural processes and increased the vulnerability of aquatic systems.

One major challenge for humans in the future will be to adapt in a way to global warming that will minimize the impact on aquatic ecosystems by minimizing pollution, introduction of exotic species into the ecosystem, habitat destruction, and further fragmentation of existing freshwater marine habitats.

The Pew Center stresses that humans greatly depend on inland freshwater and coastal wetland ecosystems to provide critical goods and services every day. If populations want that to continue, then scientists and policy-makers must focus more on minimizing negative effects to freshwater ecosystems and educating people about the best ways to maintain the health and longevity of these precious natural resources in a changing world.

Conclusions—Where to Go from Here

As global warming progresses and temperatures continue to rise, the effects on the Earth's ecosystems will continue to intensify. Climatologists have developed general circulation models (GCMs) and can enter data into them—such as carbon dioxide (CO_2) levels—adjusting and controlling the variables to see what will happen to the ecosystems being studied. For example, they can adjust the levels of CO_2 and hold all the other variables in the model constant to see just what effect CO_2 has at different concentrations. Mathematical models coupled with a sound understanding of ecological and biological concepts are what climatologists must rely on to predict, plan, and attempt to manage for the future. This final chapter looks at the process of adaptation and the realities that species must face under a rapidly warming world. It then looks at some of the conclusions climatologists have come to concerning global climate change and the important lessons learned, as well as what the crucial considerations are for the future. It concludes with the identification of ways that everyone can pitch in and help ease the situation right now for the unfortunate wildlife being affected by rising temperatures that are causing their habitats to disappear, as well as ways humans can provide for and protect the environment today and in the future.

ADAPTATION

In order for species to survive the effects of global warming, they must be able to evolve, or adapt. For a species to make a major adaptation (such as a response to temperature or oxygen level), it takes a long time. Adaptations can take thousands, tens of thousands, or even more years to occur successfully. At the rate global warming is progressing, species generally cannot physiologically adapt fast enough. Throughout history, organisms have evolved mechanisms to adjust to the climatic conditions under which they are able to live and reproduce. Biologically, in response to a change in climate they will adapt autonomously—naturally on their own. But because species must depend on their own natural internal ability to adapt, the adaptive processes cannot be sped up just because the changes in the environment that are making the changes necessary have unnaturally sped up. The harsh reality is this: If suitable habitat conditions disappear or shift in position at a faster rate than the affected species can adjust or migrate, extinctions will occur. The unfortunate result is that there will be less diversity in the ecosystems, making them less strong and healthy. In this situation, many of the roles that the previous, more diverse ecosystem played will cease to exist, ultimately harming the entire ecosystem.

According to Dr. Jay R. Malcolm of the University of Toronto and Dr. Louis F. Pitelka of the University of Maryland, who prepared a climate change report for the Pew Center on Global Climate Change, it is not evolutionary change that will save species over the next few centuries. They say the only types of species that true evolutionary change could help are those with very short life cycles (a year or less). Unfortunately, the species that fit into this class are the weedy plants and pests, and they would probably be able to evolve.

Therefore, these experts have recommended a different type of solution to the problem. In order for species to remain as strong and healthy as possible, the best solution is to plan ahead and maintain their overall ecosystem structure and species composition. They have identified the following four ways to do this:

1. Reduce fragmentation (keep the species together, not spread out all over the landscape).
2. Keep current habitats from becoming degraded (keep invasive species and pollution out, control erosion and other land degradation processes).
3. Increase the connectivity among habitat blocks and fragments (create corridors between habitat areas so that species can maintain contact and freely move about with each other).
4. Reduce external anthropogenic environmental stresses (keep urbanization and other human activity away).

In order to make this successful, land and resource managers must already have existing strategies in place to conserve biodiversity and protect natural resources. It should not be a quick plan put in place at the last minute as an emergency response to a warming climate. Plans should already be well thought out ahead of time, taking into account all of the cause-and-effect variables. Issues that need to be taken into account include designating sufficient areas for species to exist in now, as well as future habitat they may need to migrate to, and appropriate planning to enable the species to move into the already identified area. Species that currently live outside of protected areas also need to be accounted for. When climate changes, migration routes of these species may need to be taken into consideration, planned out, and provided for. Environments intact presently need to be kept healthy and unpolluted, enabling species to remain healthy and strong, so that when it comes time to move, they will be able to. In situations where species cannot adapt in time, humans must plan for the health and well-being of ecosystems in order to preserve them for the future.

CONCLUSIONS DRAWN/LESSONS LEARNED

Through studies of paleoclimatic data, it has emerged that climate is one of the most important factors in controlling the distribution and survival of vegetation and animal species. It has usually been variations in climate that have caused shifts in the distribution of species and changes in the health of ecosystems.

Based on models that have been created by several scientists at the National Aeronautics and Space Administration (NASA) and used by the Intergovernmental Panel on Climate Change (IPCC), notwithstanding slight differences in their outcomes, they do agree that global warming will cause major shifts in ecosystems as it forces species to shift poleward in latitude and in elevation up mountain ranges toward cooler temperatures. Some changes in ecosystems will be so great that not all species will survive, ultimately creating a less diverse world. The faster and more severe the shift, the more likely that there will be a reduction in species diversity. This will be a classic example of survival of the fittest.

Changing ecosystems will radically affect all aspects of our world. For example, agricultural production may shift northward into Canada from the Great Plains of the United States. Ski resorts in Colorado, Utah, and California may no longer have a reliable amount of snow to run successful winter resort businesses. The Winter Olympics traveling to different countries around the world every four years may become a thing of the past. The Earth's brightly colored coral reefs may eventually appear just a ghostly white—a haunting reminder of cooler, better, healthier days gone by.

Because climate models are not yet to the point where they can accurately project what will happen to small areas—such as specific bays, drainage basins, mountain ranges, or urban environments—there is still quite a bit of uncertainty involved, making it difficult to create successful, efficient climate management plans. Because humans have never had to deal with global warming before now, the research and discoveries being made today are like conducting a mammoth science experiment on an Earth-sized laboratory.

Understanding global warming is also not a clear issue. Because there are other natural factors that contribute to the Earth's climate (as discussed earlier in this book) it is often difficult to determine how much of the changing climate is due to global warming and how much is not. This represents a learning opportunity for every person on Earth.

SPECIAL REPORT GLOBAL WARMING

TIME

BE WORRIED. BE VERY WORRIED.

Climate change isn't some vague
future problem—it's already
damaging the planet at an alarming
pace. Here's how it affects you,
your kids and their kids as well

EARTH AT THE TIPPING POINT

HOW IT THREATENS YOUR HEALTH

**HOW CHINA & INDIA CAN HELP
SAVE THE WORLD—OR DESTROY IT**

THE CLIMATE CRUSADERS

Global warming is not a situation waiting to happen in the future.
It is a very real condition that is negatively affecting and changing
ecosystems right now. If its progression is not slowed and stopped,
damage to ecosystems will be irreversible and species will no longer
be able to survive. *(Nature's Images)*

Global warming is so much more than a scientific issue that is up
for discussion. It has proven to be genuine, serious, advancing at an
alarming rate, and involving every person on Earth, unlike anything
else that has come before it. For this reason, humans must also be able
to adapt—accept changes in energy sources, harvest food differently,

adjust to shifts in the availability of goods and services, and alter recreation, lifestyle, and many other aspects of their personal lives. Global warming is a real problem that will touch everybody on Earth—no one will escape it untouched.

WAYS TO HELP

Many look at global warming as the "problem that someone out there needs to start doing something about," but they do not stop and ask "and just who exactly would that someone be"? The reality is that all humans on Earth are part of the problem to one degree or another, and all need to do their part to save the Earth. An important point to stress here is that not all is lost—climatologists at Defenders of Wildlife believe that global warming is a problem that we can still solve. If action is taken immediately by all countries, the global warming progression could be slowed down significantly. In this age of technology and new scientific breakthroughs happening all the time, finding a solution looks promising.

The future of the world's wildlife and plants is at the mercy of the actions humans make today. Our future needs to be under the guidance of qualified scientists and policy-makers. If plans are put into effect that deal with present behaviors, such as changing to renewable energy instead of burning fossil fuels, conservation of natural resources, and respect for the environment—even though it may involve changing personal lifestyle choices—humans are resilient, determined, and can succeed. Although there are already enough greenhouse gases in the atmosphere that some global warming is going to happen regardless of what actions are taken from this time forward, if everyone works toward a common goal, the battle over global warming will eventually be won and the Earth's ecosystems will be able to breathe a little easier.

CHRONOLOGY

ca. 1400–1850   Little Ice Age covers the Earth with record cold, large glaciers, and snow. There is widespread disease, starvation, and death.

1800–70   The levels of CO_2 in the atmosphere are 290 ppm.

1824   Jean-Baptiste Joseph Fourier, a French mathematician and physicist, calculates that the Earth would be much colder without its protective atmosphere.

1827   Jean-Baptiste Joseph Fourier presents his theory about the Earth's warming. At this time many believe warming is a positive thing.

1859   John Tyndall, an Irish physicist, discovers that some gases exist in the atmosphere that block infrared radiation. He presents the concept that changes in the concentration of atmospheric gases could cause the climate to change.

1894   Beginning of the industrial pollution of the environment.

1913–14   Svante Arrhenius discovers the greenhouse effect and predicts that the Earth's atmosphere will continue to warm. He predicts that the atmosphere will not reach dangerous levels for thousands of years, so his theory is not received with any urgency.

1920–25   Texas and the Persian Gulf bring productive oil wells into operation, which begins the world's dependency on a relatively inexpensive form of energy.

1934   The worst dust storm of the dust bowl occurs in the United States on what historians would later call Black Sunday. Dust storms are a product of drought and soil erosion.

1945   The U.S. Office of Naval Research begins supporting many fields of science, including those that deal with climate change issues.

1949–50 Guy S. Callendar, a British steam engineer and inventor, propounds the theory that the greenhouse effect is linked to human actions and will cause problems. No one takes him too seriously, but scientists do begin to develop new ways to measure climate.

1950–70 Technological developments enable increased awareness about global warming and the enhanced greenhouse effect. Studies confirm a steadily rising CO_2 level. The public begins to notice and becomes concerned with air pollution issues.

1958 U.S. scientist Charles David Keeling of the Scripps Institution of Oceanography detects a yearly rise in atmospheric CO_2. He begins collecting continuous CO_2 readings at an observatory on Mauna Loa, Hawaii. The results became known as the famous Keeling Curve.

1963 Studies show that water vapor plays a significant part in making the climate sensitive to changes in CO_2 levels.

1968 Studies reveal the potential collapse of the Antarctic ice sheet, which would raise sea levels to dangerous heights, causing damage to places worldwide.

1972 Studies with ice cores reveal large climate shifts in the past.

1974 Significant drought and other unusual weather phenomenon over the past two years cause increased concern about climate change not only among scientists but with the public as a whole.

1976 Deforestation and other impacts on the ecosystem start to receive attention as major issues in the future of the world's climate.

1977 The scientific community begins focusing on global warming as a serious threat needing to be addressed within the next century.

1979 The World Climate Research Programme is launched to coordinate international research on global warming and climate change.

1982 Greenland ice cores show significant temperature oscillations over the past century.

1983 The greenhouse effect and related issues get pushed into the political arena through reports from the U.S. National Academy of Sciences and the Environmental Protection Agency.

1984–90 The media begins to make global warming and its enhanced greenhouse effect a common topic among Americans. Many critics emerge.

1987 An ice core from Antarctica analyzed by French and Russian scientists reveals an extremely close correlation between CO_2 and temperature going back more than 100,000 years.

1988 The United Nations set up a scientific authority to review the evidence on global warming. It is called the Intergovernmental Panel on Climate Change (IPCC) and consists of 2,500 scientists from countries around the world.

1989 The first IPCC report says that levels of human-made greenhouse gases are steadily increasing in the atmosphere and predicts that they will cause global warming.

1990 An appeal signed by 49 Nobel prizewinners and 700 members of the National Academy of Sciences states, "There is broad agreement within the scientific community that amplification of the Earth's natural greenhouse effect by the buildup of various gases introduced by human activity has the potential to produce dramatic changes in climate . . . Only by taking action now can we insure that future generations will not be put at risk."

1992 The United Nations Conference on Environment and Development (UNCED), known informally as the Earth Summit, begins on June 3 in Rio de Janeiro, Brazil. It results in the United Nations Framework Convention on Climate Change, Agenda 21, the Rio Declaration on Environment and Development Statement of Forest Principles, and the United Nations Convention on Biological Diversity.

1993 Greenland ice cores suggest that significant climate change can occur within one decade.

1995 The second IPCC report is issued and concludes there is a human-caused component to the greenhouse effect warming. The consensus is that serious warming is likely in the coming century. Reports on the breaking up of Antarctic ice sheets and other signs of warming in the polar regions are now beginning to catch the public's attention.

1997 The third conference of the parties to the Framework Convention on Climate Change is held in Kyoto, Japan. Adopted on December 11, a document called the Kyoto Protocol commits its signatories to reduce emissions of greenhouse gases.

2000 Climatologists label the 1990s the hottest decade on record.

2001 The IPPC's third report states that the evidence for anthropogenic global warming is incontrovertible, but that its effects on climate are still difficult to pin down. President Bush declares scientific uncertainty too great to justify Kyoto Protocol's targets.

The United States Global Change Research Program releases the findings of the National Assessment of the Potential Consequences of Climate Variability and Change. The assessment finds that temperatures in the United States will rise by 5 to 9°F (3–5°C) over the next century and predicts increases in both very wet (flooding) and very dry (drought) conditions. Many ecosystems are vulnerable to climate change. Water supply for human consumption and irrigation is at risk due to increased probability of drought, reduced snow pack, and increased risk of flooding. Sea-level rise and storm surges will most likely damage coastal infrastructure.

2002 Second hottest year on record.

Heavy rains cause disastrous flooding in Central Europe leading to more than 100 deaths and more than $30 billion in damage. Extreme drought in many parts of the world (Africa, India,

Australia, and the United States) results in thousands of deaths and significant crop damage. President Bush calls for 10 more years of research on climate change to clear up remaining uncertainties and proposes only voluntary measures to mitigate climate change until 2012.

2003   U.S. senators John McCain and Joseph Lieberman introduce a bipartisan bill to reduce emissions of greenhouse gases nation-wide via a greenhouse gas emission cap and trade program.

Scientific observations raise concern that the collapse of ice sheets in Antarctica and Greenland can raise sea levels faster than previously thought.

A deadly summer heat wave in Europe convinces many in Europe of the urgency of controlling global warming but does not equally capture the attention of those living in the United States.

International Energy Agency (IEA) identifies China as the world's second largest carbon emitter because of their increased use of fossil fuels.

The level of CO_2 in the atmosphere reaches 382 ppm.

2004   Books and movies feature global warming.

2005   Kyoto Protocol takes effect on February 16. In addition, global warming is a topic at the G8 summit in Gleneagles, Scotland, where country leaders in attendance recognize climate change as a serious, long-term challenge.

Hurricane Katrina forces the U.S. public to face the issue of global warming.

2006   Former U.S. vice president Al Gore's *An Inconvenient Truth* draws attention to global warming in the United States.

Sir Nicholas Stern, former World Bank economist, reports that global warming will cost up to 20 percent of worldwide gross domestic product if nothing is done about it now.

2007   IPCC's fourth assessment report says glacial shrinkage, ice loss, and permafrost retreat are all signs that climate change is underway now. They predict a higher risk of drought, floods, and more powerful storms during the next 100 years. As a result, hunger, homelessness, and disease will increase. The atmosphere may warm 1.8 to 4.0°C and sea levels may rise 7 to 23 inches (18 to 59 cm) by the year 2100.

Al Gore and the IPCC share the Nobel Peace Prize for their efforts to bring the critical issues of global warming to the world's attention.

2008   The price of oil reached and surpassed $100 per barrel, leaving some countries paying more than $10 per gallon.

Energy Star appliance sales have nearly doubled. Energy Star is a U.S. government-backed program helping businesses and individuals protect the environment through superior energy efficiency.

U.S. wind energy capacity reaches 10,000 megawatts, which is enough to power 2.5 million homes.

2009   President Obama takes office and vows to address the issue of global warming and climate change by allowing individual states to move forward in controlling greenhouse gas emissions. As a result, American automakers can prepare for the future and build cars of tomorrow and reduce the country's dependence on foreign oil. Perhaps these measures will help restore national security and the health of the planet, and the U.S. government will no longer ignore the scientific facts.

The year 2009 will be a crucial year in the effort to address climate change. The meeting on December 7–18 in Copenhagen, Denmark, of the UN Climate Change Conference promises to shape an effective response to climate change. The snapping of an ice bridge in April 2009 linking the Wilkins Ice Shelf (the size of Jamaica) to Antarctic islands could cause the ice shelf to break away, the latest indication that there is no time to lose in addressing global warming.

# GLOSSARY

adaptation   an adjustment in natural or human systems to a new or changing environment; adaptation to climate change refers to adjustment in natural or human systems in response to actual or expected climate changes.

aerosols   tiny bits of liquid or solid matter suspended in air that come from natural sources such as erupting volcanoes and from waste gases emitted from automobiles, factories, and power plants; by reflecting sunlight, they cool the climate and offset some of the warming caused by greenhouse gases.

albedo   the relative reflectivity of a surface; a surface with high albedo reflects most of the light that shines on it and absorbs very little energy; a surface with a low albedo absorbs most of the light that shines on it and reflects very little.

anthropogenic   resulting from human activities; usually used in the context of emissions that are produced as a result of human activities.

atmosphere   the thin layer of gases that surround the Earth and allow living organisms to breathe, it reaches 400 miles (644 km) above the surface, but 80 percent is concentrated in the troposphere—the lower seven miles (11 km) above the Earth's surface.

baculoviruses   a family of large rod-shaped viruses that can be divided into two genera: nucleopolyhedroviruses (NPV) and granuloviruses (GV). Baculovirus-produced proteins are currently being researched as therapeutic cancer vaccines. These viruses infect and kill a large number of different invertebrate species, especially insects. They are used as biocontrol agents.

biodiversity   different plant and animal species

biomass   plant material that can be used for fuel

bleaching (coral)   the loss of algae from corals that causes the corals to turn white; one of the results of global warming, signifying a die-off of unhealthy coral.

carbon   a naturally abundant nonmetallic element that occurs in many inorganic and in all organic compounds.

carbon cycle   a colorless, odorless gas that forms when carbon atoms combine with oxygen atoms; a tiny, but vital, part of the atmosphere. The heat-absorbing ability of CO_2 is what makes life possible on Earth.

carbon dioxide   a colorless, odorless gas that passes out of the lungs during respiration. It is the primary greenhouse gas and causes the greatest amount of global warming.

carbon sink   an area where large quantities of carbon are built up in the wood of trees, in calcium carbonate rocks, in animal species, in the ocean, or any other place where carbon is stored. These places act as reservoirs, keeping carbon out of the atmosphere.

climate   the usual pattern of weather that is averaged over a long period of time.

climate model   a quantitative way of representing the interactions of the atmosphere, oceans, land surface, and ice. Models can range from relatively simple to extremely complicated.

climatologist   a scientist who studies the climate.

concentration   the amount of a component in a given area or volume. In global warming, it is a measurement of how much of a particular gas is in the atmosphere compared to all of the gases in the atmosphere.

condense   the process that changes a gas into a liquid.

conterminous   having a common boundary without a gap; an example being the 48 United States (all except Alaska and Hawaii).

deforestation   the large-scale cutting of trees from a forested area, often leaving large areas bare and susceptible to erosion.

desertification   the process that turns an area into a desert.

ecological   the protection of the air, water, and other natural resources from pollution or its effects. It is the practice of good environmentalism.

ecosystem   a community of interacting organisms and their physical environment.

El Niño   a cyclic weather event in which the waters of the eastern Pacific Ocean off the coast of South America become much warmer than normal and disturb weather patterns across the region. Every few years, the temperature of the western Pacific rises several degrees above that of waters to the east. The warmer water moves eastward, causing shifts in ocean currents, jet-stream winds, and weather in both the Northern and Southern Hemispheres.

emissions   the release of a substance (usually a gas when referring to the subject of climate change) into the atmosphere.

estuary   the widened tidal mouth of a river valley where freshwater comes into contact with seawater and where tidal effects are evident.

evaporation   the process by which a liquid, such as water, is changed to a gas.

evapotranspiration   the transfer of moisture from the Earth to the atmosphere by evaporation of water and transpiration from plants.

feedback   a change caused by a process that, in turn, may influence that process. Some changes caused by global warming may hasten the process of warming (positive feedback); some may slow warming (negative feedback).

food chain   the transfer of food energy from producer (plant) to consumer (animal) to decomposer (insect, fungus, etc.).

forb   an herbaceous plant that is not a grass.

fossil fuel   an energy source made from coal, oil, or natural gas. The burning of fossil fuels is one of the chief causes of global warming.

glacier   a mass of ice formed by the buildup of snow over hundreds and thousands of years.

global warming   an increase in the temperature of the Earth's atmosphere, caused by the buildup of greenhouse gases. This is also referred to as the "enhanced greenhouse effect" caused by humans.

great ocean conveyor belt   a global current system in the ocean that transports heat from one area to another.

greenhouse effect   the natural trapping of heat energy by gases present in the atmosphere, such as CO_2, methane, and water vapor. The trapped heat is then emitted as heat back to the Earth.

greenhouse gas   a gas that traps heat in the atmosphere and keeps the Earth warm enough to allow life to exist.

Gulf Stream   a warm current that flows from the Gulf of Mexico across the Atlantic Ocean to northern Europe. It is largely responsible for Europe's milder climate.

hydrologic cycle   the natural sequence through which water passes into the atmosphere as water vapor, precipitates to earth in liquid or solid form, and ultimately returns to the atmosphere through evaporation.

Industrial Revolution   the period during which industry developed rapidly as a result of advances in technology. This took place in Britain during the late 18th and early 19th centuries.

Intergovernmental Panel on Climate Change (IPCC)   This is an organization of 2,500 scientists that assesses information in the scientific and technical literature related to the issue of climate change. The United Nations Environment Programme and the World Meteorological Organization established the IPCC jointly in 1988.

land use   the management practice of a certain land cover type. Land use may be such things as forest, arable land, grassland, urban land, and wilderness.

land-use change   an alteration of the management practice on a certain land cover type. Land-use changes may influence climate systems because they impact evapotranspiration and sources and sinks of greenhouse gases. An example of land-use change is cutting down a forest to build a city.

methane   a colorless, odorless, flammable gas that is the major ingredient of natural gas. Methane is produced wherever decay occurs and little or no oxygen is present.

niche   an environment in which an organism could successfully survive and reproduce in the absence of competition.

nitrogen   as a gas, nitrogen takes up 80 percent of the volume of the Earth's atmosphere; also an element in substances such as fertilizer.

nitrous oxide   a heat-absorbing gas in the Earth's atmosphere; emitted from nitrogen-based fertilizers.

ozone   a molecule that consists of three oxygen atoms. Ozone is present in small amounts in the Earth's atmosphere at 14 to 19 miles (23–31 km) above the Earth's surface. A layer of ozone makes life possible by shielding the Earth's surface from most harmful ultraviolet rays. In the lower atmosphere, ozone emitted from auto exhausts and factories is an air pollutant.

parts per million (ppm)   the number of parts of a chemical found in one million parts of a particular gas, liquid, or solid.

permafrost   permanently frozen ground in the Arctic. As global warming increases, this ground is melting.

photosynthesis   the process by which plants make food using light energy, carbon dioxide, and water.

protocol   the terms of a treaty that have been agreed to and signed by all parties.

proxies   methods of determining values such as temperatures and rainfall by using substitutes, which give indirect measurements. Tree rings serve as proxies for determining rainfall abundance.

radiation   the particles or waves of energy.

renewable   something that can be replaced or regrown, such as trees, or a source of energy that never runs out, such as solar energy, wind energy, or geothermal energy.

resources   the raw materials from the Earth that are used by humans to make useful things.

rotation   the movement or path of the Earth, turning on its axis.

satellite   any small object that orbits a larger one. Artificial satellites carry instruments for scientific study and communication. Imagery taken from satellites is used to monitor aspects of global warming such as glacier retreat, ice cap melting, desertification, erosion, hurricane damage, and flooding. Sea-surface temperatures and measurements are also obtained from man-made satellites in orbit around the Earth.

simulation   a computer model of a process that is based on actual facts. The model attempts to mimic, or replicate, actual physical processes.

symbiosis   the relationship that exists between two different organisms that live in close association, with at least one being helped without either being harmed.

symbiotic   a relationship where two dissimilar organisms share a mutually beneficial relationship.

temperate   an area that has a mild climate and different seasons.

thermal   something that relates to heat.

tropical   a region that is hot and often wet (humid). These areas are located around the Earth's equator.

tropical storm   a cyclonic storm having winds ranging from approximately 30 to 75 miles (48 to 121 kilometers) per hour.

tundra   a vast treeless plain in the Arctic with a marshy surface covering a permafrost layer.

upwelling   the process by which warm, less-dense surface water is drawn away from along a shore by offshore currents and replaced by cold, denser water brought up from the subsurface.

weather   the conditions of the atmosphere at a particular time and place. Weather includes such measurements as temperature, precipitation, air pressure, and wind speed and direction.

FURTHER RESOURCES

BOOKS

Christianson, Gale. *Greenhouse: The 200-Year Story of Global Warming.* New York: Walker, 1999. This book looks at the enhanced greenhouse effect worldwide after the Industrial Revolution and outlines the consequences to the environment.

Cox, John D. *Climate Crash—Abrupt Climate Change and What It Means for Our Future.* Washington, D.C.: Joseph Henry Press, 2005. This book examines the science of paleoclimatology and how events of the past hold clues to the present and future.

Flannery, Tim. *The Weather Makers.* New York: Atlantic Monthly Press, 2005. This book explores how humans are changing the climate and how it will affect all life on Earth.

Friedman, Katherine. *What If the Polar Ice Caps Melted?* Danbury, Conn.: Children's Press, 2002. This book focuses on environmental problems related to the Earth's atmosphere, including global warming, changing weather patterns, and their effects on ecosystems.

Gelbspan, Ross. *The Heat Is On: The High Stakes Battle over Earth's Threatened Climate.* Reading, Mass.: Addison Wesley, 1997. This work offers a look at the controversy environmentalists often face when they deal with fossil fuel companies.

———. *Boiling Point.* New York: Basic Books, 2004. This publication presents the role that politicians, oil companies, the media, and activists have had in shaping people's beliefs about the issue of global warming.

Gore, Al. *An Inconvenient Truth.* Emmaus, Penn.: Rodale, 2006. This publication presents an excellent overview of the global warming problem, how it has come about, what it means for the future, and why humans need to act now to slow it down.

Hamblin, W. Kenneth, and Eric H. Christiansen. *Earth's Dynamic Systems.* 10th ed. Upper Saddle River, N.J.: Prentice Hall, 2000. This book discusses the physical aspects of Earth science, explaining the physical mechanisms behind plate tectonics, glaciers, shorelines, groundwater, river systems, and other natural systems that are affected by global warming.

Harrison, Patrick 'GB', Gail 'Bunny' McLeod, and Patrick G. Harrison. *Who Says Kids Can't Fight Global Warming.* Chattanooga, Tenn.: Pat's Top Products, 2007. This book offers real solutions that everybody can do to help solve the world's biggest air pollution problems.

Hassol, Susan Joy. "Impacts of a Warming Arctic—Arctic Climate Impact Assessment." Arctic Research Center. Cambridge University Press, 2004. This takes a detailed look at the changes facing the arctic ecosystem as temperatures rise.

Houghton, John. *Global Warming: The Complete Briefing.* New York: Cambridge University Press, 2004. This book outlines the scientific basis of global warming and describes the impacts that climate change will have on society. It also looks at solutions to the problem.

Kattenberg A., et al. *"Climate Models—Projections of Future Climate. Climate Change 1995: The Science of Climate Change."* Cambridge: Cambridge University Press, 1996. This publication looks at the issue of heat waves and the effects it may have worldwide.

Langholz, Jeffrey. *You Can Prevent Global Warming (and Save Money!): 51 Easy Ways.* Riverside, N.J.: Andrews McMeel Publishing, 2003. This book aims at converting public concern over global warming into positive action to stop it by providing simple, everyday practices that can easily be done to minimize it, as well as save the person money.

Leggett, Jeremy. *The Carbon War.* New York: Routledge, 2001. This book presents the political view of global warming and the conflict with the oil industry.

Linden, Eugene. *The Winds of Change.* New York: Simon & Schuster, 2006. This book looks at past and present climate change and what that may indicate for the future.

Lynas, Mark. *High Tide.* New York: Picador, 2004. This publication explores the global effects of climate change on the Earth's diverse ecosystems.

McKibben, Bill. *Fight Global Warming Now: The Handbook for Taking Action in Your Community.* New York: Holt Paperbacks, 2007. This book provides the facts of what must change to save the climate. It also shows how everyone can act proactively in their community to make a difference.

Michaels, Patrick J. *Meltdown.* Washington, D.C: Cato Institute, 2004. This publication discusses the issues of climate change as seen not only by scientists but by politicians and the media as well.

———. *Shattered Consensus.* New York: Rowman & Littlefield Publishers, Inc., 2005. This book presents a series of essays from climate change experts on key issues surrounding global warming and interpretations of how severe the problem is.

Millennium Ecosystem Assessment. "Ecosystems and Human Well-being: Desertification Syntheses," 2005. This discusses solutions to living with desertification.

National Assessment Synthesis Team (NAST). *"Climate Change Impacts on the United States: The Potential Consequences of Climate Variability and Change, Report for the U.S. Global Change Research Program."* Cambridge: Cambridge University Press, 2001. Provides an overview on impacts due to global warming.

National Research Council. *Abrupt Climate Change—Inevitable Surprises.* Washington, D.C.: National Academy Press, 2002. This book explores the causes and reality of abrupt climate change and its global effects.

Pearce, Fred. *When the Rivers Run Dry.* Boston: Beacon Press, 2006. This book discusses the problem of drought as one of the major crises that will occur in the 21st century as a result of global warming.

Poff, N. LeRoy, Mark M. Brinson, and John W. Day, Jr. *Aquatic Ecosystems and Global Climate Change: Potential Impacts on Inland Freshwater and Coastal Wetland Ecosystems in the United States.* Prepared for the Pew Center on Global Climate Change. January, 2002. This report deals with the ecological ramifications of global warming on marine ecosystems.

Pringle, Laurence. *Global Warming: The Threat of Earth's Changing Climate.* New York: SeaStar Books, 2001. This book provides information on the carbon cycle, rising sea levels, El Niño, aerosols, smog, flooding, and other issues related to global warming.

Soloman, A. *The Nature of Past, Present, and Future Boreal Forests: Lessons for a Research and Modeling Agenda, Systems Analysis of the Global Boreal Forest.* Cambridge: Cambridge University Press, 1992. This book presents the challenges the boreal forests are facing.

Stern, Nicholas. *The Economics of Climate Change.* New York: Cambridge University Press, 2007. This book discusses the economic ramifications of global warming.

Stocker, T. F., R. Knutti, and G. K. Plattner. *The Oceans and Rapid Climate Change: Past, Present, and Future.* Washington, D.C.: American Geophysical Union, 2001. This presents a perspective on the future of the thermohaline circulation.

Thornhill, Jan. *This Is My Planet—The Kids Guide to Global Warming.* Toronto, Ontario: Maple Tree Press, 2007. This book offers students the tools they need to become ecologically oriented by taking a comprehensive look at climate change in polar, ocean, and land-based ecosystems.

University of California, Davis, News and Information. "Rangeland's Role in Buffering Global Warming Explored." December 16, 1997. This presents the issue of rangeland and global warming.

U.S. Climate Change Science Program/U.S. Global Change Research Programme. "U.S. National Assessment of Climate Change." Washington, D.C., 2004. Provides a good overview of relevant issues of global warming.

Weart, Spencer R. *The Discovery of Global Warming (New Histories of Science, Technology, and Medicine).* Cambridge, Mass.: Harvard University Press, 2004. Traces the history of the global warming concept through a long process of incremental research rather than a dramatic revelation.

World Resources Institute (WRI). "People and Ecosystems: The Fraying Web of Life." Prepared by the United Nations Development Programme (UNDP), The World Bank, and the World Resources Institute. World Resources 2000–2001. This shows the close interaction that exists between people and their environments.

World Wildlife Fund. "New Report Shows Global Warming Link to Australia's Worst Drought." January 13, 2003. Discusses one of Australia's worst droughts.

JOURNAL ARTICLES

Alley, Richard B. "Abrupt Climate Change." *Scientific American.* (November 2004): 62–69. This article discusses the great ocean conveyor belt current and how global warming could shut it down and trigger another ice age.

Appenzeller, Tim. "The Big Thaw." *National Geographic.* (June 2007): 58–71. This article gives an overview of the global melting of ice from glaciers and ice caps and what that means for the future.

Applebome, Peter. "A Community Tries to Shrink Its Footprint." *New York Times* (1/20/08). Available online. URL: http://www.nytimes.com/2008/01/20/nyregion/20towns.html. Accessed June 28, 2009. This article profiles Westport, Connecticut, and the steps they are taking to become more environmentally responsible.

Astor, Michael. "Warming May Change Amazon." Associated Press (12/29/06). Available online. URL: http://www.washingtonpost.com/wp-dyn/content/article/2006/12/29/AR2006122901144.html. Accessed June 28, 2009. This looks at the effects of global warming in the world's largest tropical rain forest.

Bachelet, D., J. Lenihan, R. Drapek, and R. Neilson. "VEMAP vs. VINCERA: A DGVM Sensitivity to Differences in Climate Scenarios." *Global and Planetary Change,* vol. 64, 1–2. (November 2008): 38–48. This article compares two models designed to predict future climate scenarios.

Barringer, Felicity. "U.S. Given Poor Marks on the Environment." *New York Times* (1/23/08). Available online. URL: http://www.nytimes.

com/2008/01/23/washington/23enviro.html. Accessed June 28, 2009. Compared to other countries, the United States lags behind on environmental responsibility and positive action.

BBC News. "Deserts Need Better Management." (6/5/06). Available online. URL: http://news.bbc.co.uk/1/hi/sci/tech/5041988.stm. Accessed June 28, 2009. This review outlines the susceptibility of desert ecosystems and what global warming is doing to them.

———. "OECD Warns of Alpine Ski Future." (12/13/06). Available online. URL: http://news.bbc.co.uk/2/hi/business/6176271.stm. Accessed January 10, 2009. This discusses the struggle ski resorts in the European Alps are now having under the effects of global warming.

Bindschadler, Robert A., and Charles R. Bentley. "On Thin Ice?" *Scientific American* (December 2002): 98–105. This publication discusses the melting rates of the Earth's most massive ice sheets and what controls the rates of disintegration.

Bloomfield, Janine, and Steven Hamburg. *Seasons of Change: Global Warming and New England's White Mountains.* Washington, D.C.: Environmental Defense Fund, 1997. This discusses the effects global warming may have on the New England region of the United States in the coming years.

Borenstein, Seth. "Hotter Weather Linked to More Extinctions." Discovery News (10/24/07). Available online. URL: http://dsc.discovery.com/news/2007/10/24/extinctions-warming-print.html. Accessed June 28, 2009. This article examines the Earth's past and how scientists study extinctions, as well as what today's global warming may have on present life on Earth.

Bosire, Bogonko. "More Than 1 Billion Trees Planted in 2007." Discovery News (11/28/07). Available online. URL: http://dsc.discovery.com/news/2007/11/28/tree-planting-forests-print.html. Accessed January 10, 2009. This article describes how more than 1 billion trees were planted around the world in 2007, with Ethiopia and Mexico leading in the drive to combat climate change.

Bourne, Joel K., Jr. "Green Dreams." *National Geographic.* (October 2007): 38–59. This article explores the use of biofuels.

Britt, Robert Roy. "Surprising Side Effects of Global Warming."
    LiveScience (12/22/04). Available online. URL: http://www.
    livescience.com/environment. Accessed January 11, 2009. This arti-
    cle addresses the damage to the landscape from melting permafrost,
    landslides, and mudslides as a result of global warming.

———. "Global Warming Likely Cause of Worst Mass Extinction
    Ever." LiveScience (1/20/05). Available online. URL: http://www.
    livescience.com/environment/050120_great-dying.html. Accessed
    January 9, 2009. This article presents scientific evidence that past
    extinctions were related to changing climate.

———. "Caution: Global Warming May Be Hazardous to Your
    Health." LiveScience (2/2005). Available online. URL: http://www.
    livescience.com/environment/050221_warming_health.html.
    Accessed January 10, 2009. This article addresses the timely issues
    of pollution and the effects on human health.

———. "Energy Imbalance Behind Global Warming." LiveScience
    (4/28/05). Available online. URL: http://www.livescience.com/envi-
    ronment_050428/solar_energy.html. Accessed June 28, 2009. This
    article looks at existing heat brought on by air pollution slowly
    warming the Earth to a point where it will soon be out of control.

———. "125 Large Northern Lakes Disappear." LiveScience (6/3/05).
    Available online. URL: http://www.livescience.com/environ-
    ment/050603_lakes_gone.html. Accessed June 28, 2009. This
    article focuses on lakes in the Arctic that have vanished as tem-
    peratures have risen over the past two decades and what effect this
    is having.

———. "Arctic Summer Could be Ice-free by 2105." LiveScience
    (8/23/05). Available online. URL: http://www.livescience.com/envi-
    ronment/050823_ice_free.html. Accessed June 28, 2009. This dis-
    cusses the impacts currently felt by the fragile Arctic ecosystem and
    the results of prolonged stress.

———. "Global Warming Sparks Increased Plant Production in Arctic
    Lakes." LiveScience (10/24/05). Available online. URL: http://www.
    livescience.com/environment/051024_arctic_lakes.html. Accessed

June 28, 2009. This article looks at longer growing seasons in the Arctic ecosystems as a result of global warming.

———. "Insurance Company Warns of Global Warming's Costs." LiveScience (11/1/05). Available online. URL: http://www. livescience.com/environment/051101_insurance_warning.html. Accessed June 28, 2009. This article explores the economic issues behind the destruction caused by global warming.

———. "The Irony of Global Warming: More Rain, Less Water." LiveScience (11/16/05). Available online. URL: http://www. livescience.com/environment/051116_water_shortage.html. Accessed June 28, 2009. This article outlines the fact that an enhanced water cycle may not benefit the necessary water supplies.

———. "Conflicting Claims on Global Warming and Why It's All Moot." LiveScience (2/1/06). Available online. URL: http://www. livescience.com/environment/060201_temperature_differences. html. Accessed August 23, 2008. This presents the pros and cons of opinions on global warming.

Carey, Bjorn. "Soot Could Hasten Melting of Arctic Ice." LiveScience (3/28/05). Available online. URL: http://www.livescience.com/envi-ronment/050328_arctic-soot.html. Accessed January 5, 2009. This article shows that when ice becomes polluted with soot, it changes the reflective properties and accelerates melting in the Arctic.

———. "Arctic Summer Could Be Ice-Free by 2105." LiveScience (8/23/05). Available online. URL: http://www.livescience.com/envi-ronment/050823_ice_free.html. Accessed June 28, 2009. Discusses the accelerated rates of ice melting.

Carlton, Jim. "Is Global Warming Killing the Polar Bears?" *Wall Street Journal* (12/14/05). Available online. URL: http://online.wsj. com/public/article_print/SB113452435089621905-vnekw47PQGt-Dyf3iv5XE N71_o5I_20061214.html. Accessed June 28, 2009. This article explores the dilemma polar bears are facing due to melting polar ice.

Castle, Stephen, and James Kanter. "Stricter System to Trim Carbon Emissions Is Considered in Europe." *New York Times* (1/22/08).

Available online. URL: http://www.nytimes.com/2008/01/22/business/worldbusiness/22emissions.html. Accessed June 28, 2009. This article explores the concept of reducing carbon emissions.

CBS News. "'Monumental' Climate Change?" (6/19/07). Available online. URL: http://www.cbsnews.com/stories/2007/06/19/evening-news/main2952285.shtml. Accessed June 30, 2009. This article outlines the destruction on the world's fine art as a result of pollution and global warming.

Chapin III, F., E. Zavaleta, et al. "Consequences of Changing Biodiversity." In *Nature* 405, 234–242. This review examines the impacts that will be felt by the environment as global warming intensifies.

Chevalier, Judith. "A Carbon Cap That Starts in Washington." *New York Times* (12/16/07). Available online. URL: http://www.nytimes.com/2007/12/16/business/16view.html. Accessed June 29, 2009. This article looks at the political side of global warming as an international issue.

Clayton, Mark. "A New Gust of Wind Projects in U.S." CBS News (1/22/06). Available online. URL: http://www.usatoday.com/money/industries/energy/2006-01-18-wind-power_x.htm. Accessed June 29, 2009. This article discusses several new wind-generated energy farms being built in Texas as a source of clean energy.

CNN, Environmental News Network. "265,000 Flee as Massive Wildfires Char Southern California." (10/22/07). Available online. URL: http://ballhype.com/story1265_000_flee_as_massive_wildfires_char_southern/. Accessed June 29, 2009. Discusses current issues of wildfires.

Coleman, Joseph. "Seaweed: The New Carbon Sink?" Discovery News (12/10/07). Available online. URL: http://dsc.discovery.com/news. Accessed December 6, 2008. This article looks at the world's carbon sinks and their potential to offset the effects of global warming.

Cook, T., M. Folli, et al. "The Relationship Between Increasing Sea-surface Temperature and the Northward Spread of *Perkinsus Marinus* (Dermo) Disease Epizootics in Oysters." In *Estuarine, Coastal, and Shelf Science* 46:587–597 (1998). This looks at the effects warming oceans have on the marine environment in the midlatitudes.

Culotta, Elizabeth. "Will Plants Profit from High CO_2?" *Science* (5/5/95). Available online. URL: http://www.sciencemag.org/cgi/content/citation/268/5211/654. Accessed June 29, 2009. This article explores the possible effects of various CO_2 levels on vegetation as a result of global warming and whether they will experience enhanced growth.

Cury, P., and C. Roy. "Optimal Environmental Window and Pelagic Fish Recruitment Success in Upwelling Areas." In *Canadian Journal of Fisheries and Aquatic Science* 46:670–680 (1989).

D'Agnese, Joseph. "Why Has Our Weather Gone Wild?" *Discover* (6/2000). Available online. URL: http://www.allanstime.com/News/weather/wild.htm. Accessed June 29, 2009. This articles focuses on the recent global changes in weather, such as shifting seasons, severe storms, droughts, heat waves, and other weather-related events, and their connection to global warming.

Davidson, Sarah. "How Global Warming Can Chill the Planet." LiveScience (12/17/04). Available online. URL: http://www.livescience.com/environment/041217_sealevel_rise.html. Accessed June 29, 2009. This article explores how the oceans' currents can lead to a cooler climate if they are disrupted through the addition of freshwater from melting ice caps and glaciers.

Dean, Cornelia. "The Preservation Predicament." *New York Times* (1/29/08). Available online. URL: http://www.nytimes.com/2008/01/29/science/earth/29habi.html?ref=science. Accessed June 29, 2009. This article looks at humans' current management of the landscape and its wildlife and how habitats are going to survive in the future.

Deutsch, Claudia H. "A Threat So Big, Academics Try Collaboration." *New York Times* (12/25/07). Available online. URL: http://www.nytimes.com/2007/12/25/business/25sustain.html. Accessed June 29, 2009. This article explores the methodology of scientists working together to deal with an immense global issue.

Dudley, N. "Potential Impacts of Climate Change on Forests." A report for WWF International, 1998. This article reviews the present and future role of forests in light of global warming and the impacts they will be faced with if changes are not made today.

Eilperin, Juliet. "Severe Hurricanes Increasing, Study Finds." *Washington Post* (9/16/05). Available online. URL: http://www.washington post.com/wp-dyn/content/article/2005/09/15/AR2005091502234. html. Accessed June 29, 2009. This study links rising sea temperatures to an increase in more destructive hurricanes.

Environmental News Network Staff. "Vicious Cycle: Global Warming Feeds Fire Potential." CNN (11/2/00). Available online. URL: http://archives.cnn.com/2000/nature/11/02/global.warming.enn/index. html. Accessed June 29, 2009. This article discusses the correlation between global warming and wildfires.

Epstein, Paul R. "Is Global Warming Harmful to Health?" *Scientific American* (August 2000): 50–57. This article explores the issues of widespread epidemics as a fallout of increased global warming.

Fountain, Henry. "Katrina's Damage to Trees May Alter Carbon Balance." *New York Times* (11/20/07). Available online. URL: www. nytimes.com/2007/11/20/science/20obtree.html. Accessed June 29, 2009. This article looks at the ecological implications of the destruction of hundreds of trees on the Gulf Coast as a result of Hurricane Katrina.

———. "More Acidic Ocean Hurts Reef Algae as Well as Corals." *New York Times* (1/8/08). Available online. URL: http://www.nytimes. com/2008/01/08/science/earth/08obalga.html. Accessed June 29, 2009. As more carbon dioxide is added to the atmosphere, the oceans are becoming acidic and scientists believe they are damaging not only coral, but also reef algae.

Fox News. "Report: Global Warming Threatens Deserts, Too." (6/6/06). Available online. URL: http://www.foxnews.com/ story/0,2933,19830700.html. Accessed June 29, 2009. Discusses the impacts currently experienced in arid environments due to global warming.

Friedman, Thomas L. "It's Too Late for Later." *New York Times* (12/16/07). Available online. URL: http://www.nytimes.com/2007/12/16/opinion/16friedman.html. Accessed June 29, 2009. The UN specialists and other leaders are warning that action must be taken immediately on global warming.

Gaften, D. J., and R. J. Ross. "Increased Summertime Heat Stress in the U.S." *Nature* 396:529–530 (1998). Touches on the realities of global warming and the effects that will only get worse as temperatures continue to climb.

Gelling, Peter. "Focus of Climate Talks Shifts to Helping Poor Countries Cope." *New York Times* (12/13/07). Available online. URL: http://www.nytimes.com/2007/12/13/world/13climate.html. Accessed June 29, 2009. This article outlines several courses of realistic adaptation for countries that will be faced with the negative effects of global warming.

Gent, P. R. "Will the North Atlantic Ocean Thermohaline Circulation Weaken During the 21st Century." *Geophysical Research Letters* 28:1023–1026 (2001). This article looks at various factors that could cause the current to slow significantly, threatening its existence.

Gibbard, Seran, Ken Caldeira, et al. "Climate Effects of Global Land Cover Change." *Journal Geophysical Research Letters* (12/8/05). Available online. URL: http://www.agu.org/pubs/crossref12005/2005GL024550.shtml. Accessed June 29, 2009. This review discusses land use and its connections with the climate.

Gibbs, Walter, and Sarah Lyall. "Gore Shares Peace Prize for Climate Change Work." *New York Times* (10/13/07). Available online. URL: http://www.nytimes.com/2007/10/13/world/13nobel.html. Accessed June 29, 2009. Former vice president Al Gore was awarded the 2007 Nobel Peace Prize, sharing it with the Intergovernmental Panel on Climate Change (a network of more than 2,500 scientists).

Goudarzi, Sara. "Allergies Getting Worse Due to Global Warming." LiveScience (11/22/05). Available online. URL: http://www.livescience.com/health/051122_allergy_rise.html. Accessed June 29, 2009. This looks at how a warmer atmosphere can increase the occurrence of allergic systems in humans.

Groisman, P. Y., and D. R. Easterling. "Variability and Trends of Pre-cipitation and Snowfall Over the United States and Canada." *Journal of Climate* 7:184–205 (1994). This article looks at future increases in severe weather as a result of global warming.

Handwerk, Brian, and Lauri Hafvenstein. "Belize Reef Die-Off Due to Climate Change?" National Geographic News (3/25/04). This looks at the correlation of whether the reef die-off is due to local factors or global warming.

Hansen, James. "Defusing the Global Warming Time Bomb." *Scientific American* (March 2004). Available online. URL: news.nationalgeographic.com/news/2003/03/0325_030325_belizereefs.html. Accessed June 29, 2009. This article discusses the widespread scale of the problem today and why immediate action is urgently needed.

Heilprin, John. "Study: Earth Is Hottest Now in 2,000 Years; Humans Responsible for Much of the Warming." *USA Today* (6/23/06). This report discusses paleoclimatology and proxy evidence, as well as correlates past evidence to today's intense storms, heat waves, and other climate conditions.

Hennessey, Kathleen. "Dairy Air: Scientists Measure Cow Gas." LiveScience (7/27/05). Available online. URL: http://www.cnr.berkeley.edu/nahg/tgbl/livestock.pdf. Accessed June 29, 2009. A review of a study that looks at agricultural practices and the connections to global warming.

Hertsgaard, Mark. "On the Front Lines of Climate Change." *Time* (4/9/07). Available online. URL: http://www.time.com/time/specials/2007/environment/article/0,28804,1602354_1596572_1604879,00.html. Accessed June 29, 2009. This article explores the concepts of human adaptation to climate change.

Hoffman, Allison and Gillian Flaccus. "Wildfires Engulf Southern California." Discovery News (10/22/07). Available online. URL: http://dsc.discovery.com/news/2007/10/22/wildfires-california-print.html. Accessed June 29, 2009. This article links wildfires with global warming and climate change.

Hoffman, Paul F., and Daniel P. Schrag. "Snowball Earth." *Scientific American* (January 2000). Vol 282, no. 1, pp. 68–75. This article discusses the Earth when it was covered with ice millions of years ago.

Holechek, Jerry. "A Growing Population, Rangelands, and the Future." *Rangelands* 23(6): pp. 39–43. (December 2001). This article takes a look at the conditions of rangelands today and how today's decisions and actions will shape their conditions in the future.

Houde, E. D., and E. S. Rutherford. "Recent Trends in Estuarine Fisheries: Predictions of Fish Production and Yield." *Estuaries* 16:161–176 (1993).

Jardine, K. "The Carbon Bomb: Climate Change and the Fate of the Northern Boreal Forests." *Greenpeace International* (1994). This review explores the fragile ecosystems of the boreal forests and the impacts they will face as climate changes and biodiversity in different geographical areas responds.

Joling, Dan. "Study: Polar Bears May Turn to Cannibalism." *USA Today* (6/13/06). Available online. URL: www.asp.usatoday.com/community/tags/topic.aspx?req-tag&pg-5948tag-canada. Accessed June 29, 2009. This article discusses how the melting polar ice is keeping polar bears from being able to hunt for food in their traditional hunting grounds.

———. "Melting Sea Ice Forcing Walruses Ashore." Discovery News (10/8/07). Available online. URL: http://earthobservatory.nasa.gov/Newsroom/archive.php?cat_id=18&m=04&y =2006. Accessed June 29, 2009. Gives a detailed look at the predicament walrus face with the current lack of sea ice.

Kahn, Joseph, and Mark Landler. "China Grabs West's Smoke-Spewing Factories." *New York Times* (12/21/07). Available online. URL: http://www.nytimes.com/2007/12/21/world/asia/21transfer.html. Accessed June 29, 2009. This article looks at the energy revolution taking place in China.

Kahn, Joseph. "Japan Urges China to Reduce Pollution." *New York Times* (12/19/07). Available online. URL: http://topics.nytimes.com/top/reference/timestopics/people/k/joseph_kahn/index.html.

Accessed January 15, 2009. This article focuses on the political implications of global warming between countries.

Kanter, James. "Europe May Ban Imports of Some Biofuel Crops." *New York Times* (1/15/08). Available online. URL: http://www.nytimes.com/2008/01/15/business/worldbusiness/15biofuel.html. Accessed January 4, 2009. This article discusses a law that would prohibit the importation of fuels derived from crops grown on certain kinds of land.

Karl, T. R., and R. W. Knight. "Secular Trends of Precipitation Amount, Frequency, and Intensity in the United States." *Bulletin of the American Meteorological Society* 79 (2): 231–241 (1998). An examination of increased precipitation and its effects under the influence of global warming.

Karl, Thomas R., Neville Nicholls, and Jonathan Gregory. "The Coming Climate." *Scientific American* (May 1997). This discusses computer models and how they interpret weather patterns of a warmer world.

Karl, Thomas, and Kevin Trenberth. "The Human Impact on Climate." *Scientific American* (December 1999). This article focuses on the disruptions people cause in the natural environment and why scientists must begin to monitor and quantify the disruptions now in order to save the future.

Kaufman, Marc. "Research Shines Some Light on Mysteries of Antarctica." *Washington Post* (2/18/07). Available online. URL: http://www.washingtonpost.com/ac2/wp-dyn/emailafriend?contentId=AR2007021701335&sent=no. Accessed June 29, 2009. This article looks at snowfall, ice melt, and wind systems and how the three components work together in Antarctica.

Kay, Jane. "A Warming World: Climate Change Report." *San Francisco Chronicle* (2/3/07). Available online. URL: http://www.ucar.edu/news/pressclips/archive/07/02.jsp. Accessed June 29, 2009. This article focuses on the new report from the IPCC and their warnings of the effects of global warming to come.

Kerr, Jennifer C. "2007 Among Warmest Years Ever in U.S." *Discovery News* (12/14/07). Available online. URL: http://dsc.discovery.

com/news/2007/12/14/warmest-year-weathcr.html. Accessed June 29, 2009. Discusses the events of 2007 and the role global warming may have played.

Kluger, Jeffrey. "A Climate of Despair: Special Report, Global Warming." *Time* (4/9/01). Available online. UR: http://www.time.com/time/classroom/glenfall2001/19.html. Accessed June 29, 2009. This article discusses the political ramifications of global warming and the need for global cooperation.

———. "Is Global Warming Fueling Katrina?" *Time* (8/29/05). Available online. URL: http://www.time.com/time/printout/0,8816,1099102,00.html. Accessed June 29, 2009. This review looks at global warming as a cause of strengthening hurricanes.

———. "Global Warming Heats Up." *Time* (3/26/06). Available online. URL: http://www.time.com/time/printout/0,8816,1176980,00.html. Discusses several environmental aspects of the environment that point to the effects of global warming.

———. "What Now?" *Time* (4/9/07). This article discusses the problem from an international aspect and how everyone can be involved in the solution of the problem.

Krauss, Clifford. "As Ethanol Takes Its First Steps, Congress Proposes a Giant Leap." *New York Times* (12/18/07). Available online. URL: http://www.nytimes.come/2007/12/18/washington/18ethanol.html. Accessed June 29, 2009. This article discusses how the U.S. Congress is on the verge of writing into law one of the most ambitious dictates ever issued to American business: to create a new industry capable of converting agricultural wastes and other plant material into automotive fuel.

KTVU, San Francisco. "Report: Global Warming Threatens World's Deserts." (6/10/06). This review looks at the world's deserts and outlines UNEP's findings and predictions for the future.

Ladle, Richard J., Paul Jepson, et al. "Dangers of Crying Wolf Over Risk of Extinctions." *Nature* 428:799 (4/22/04). This article discusses the responsibility of scientists and the media to report discoveries correctly and avoid media sensationalism.

Lean, Geoffrey, and Robert Mendick. "Whale Population Devastated by Warming Oceans, Scientists Say." National Geographic News (August 2001). Available online. URL: http://news.national geographic.com/news/2001/08/0801_wirewhales2.html. Accessed June 29, 2009. Tracks the effects of global warming on whales and what that means for the future.

Lemonick, Michael D. "Life in the Greenhouse." *Time* (4/9/01). Available online. URL: http://www/time.come/time/magazine/article/0,9171,104617,00.htm. Accessed June 29, 2009. Outlines the current signs of global warming and the effects it will have on the environment.

———. "Why you Can't Ignore the Changing Climate." Parade.com (6/25/06). Available online. URL: http://www.parade.com/articles/editions/2006/edition_06-25-2006. Accessed January 19, 2009. This article points out that evidence of climate change is everywhere and not only needs to be taken seriously, but everyone needs to do their part to help fix it.

LiveScience. "Global Warming Could Overwhelm Storm Drains." (10/11/05). Available online. URL: http://www.livescience.com/environment/051011_culverts.html. Accessed June 29, 2009. This article looks at the social and management aspects of flooding during extreme storm events.

———. "Darker Days in China as Sun Gets Dimmer." (1/20/06). Available online. URL: http://www.livescience.com/environment/060119_dark_china.html. Accessed June 29, 2009. This article discusses the hazards of increased fossil fuel use and pollution.

Lovgren, Stefan. "Global Warming May Unleash 'Sand Seas' in Africa, Model Shows." National Geographic News (6/29/05). Available online. URL: http://news.nationalgeographic.com/news/2005/06/0629_050629_dunes.html. Accessed June 29, 2009. This article looks at the issue of sand dunes in Kalahari and their role in desertification as global warming intensifies.

Marshall, Carolyn. "San Francisco Fleet Is All Biodiesel." *New York Times* (12/2/07). Available online. URL: http://www.nytimes.

com/2007/12/02/us/02diesel.html. Accessed June 29, 2009. This article discusses how the city of San Francisco has converted its entire array of diesel vehicles to operate on biodiesel in order to reduce greenhouse gas emissions.

McCarthy, Michael. "The Century of Drought." *The Independent* (10/4/06). Available online. URL: http://www.commondreams.org/ headlines06/1004-02.htm. Accessed June 29, 2009. Explores the very real effects of drought and what that means for the environment.

———. "Carbon's New Math." *National Geographic* (October 2007). This article explores ways to use technology to reduce the effects of global warming.

Mouawad, Jad. "OPEC Gathering Finds High Oil Prices More Worrisome Than Welcome." *New York Times* (11/17/07). Available online. URL: http://www.nytimes.com/2007/11/17/business/17opec.html. Accessed June 29, 2009. This article discusses the economic realities of climbing oil prices and issues concerning a global economic recession.

———. "Oil Demand, the Climate and the Energy Ladder." *New York Times* (1/19/08). Available online. URL: http://www.nytimes. com/2008/01/19/business/19interview.html. Accessed June 29, 2009. An article discussing the future energy demands and the need for setting limits on carbon emissions.

National Geographic News. "Mountain Ecosystems in Danger Worldwide, UN Says." (2/1/02). Available online. URL: http://news.nationalgeographic.com/news/2002/02/0201_020201_wiremountain. html. Accessed June 29, 2009. This article addresses water shortages brought on by global warming.

Natural Resources Canada. "Climate Change Impacts and Adaptation: A Canadian Perspective." Prepared by the Climate Change Impacts and Adaptation Directorate (2002). Available online. URL: http:// adaptation.nrcan.gc.ca/perspective/index_e.php. Accessed June 29, 2009. This article discusses global warming impacts on forests.

Neergaard, Lauran. "Greenhouse Gas at 650,000-Year High." LiveScience (11/25/05). Available online. URL: http://www.

compositelive.com/environment/ap_051125_greenhouse_gas.html. Accessed June 29, 2009. This article looks at the European Project for Ice Coring in Antarctica and relates evidence found by scientists in the ice cores to the conditions humans face today.

Nicklen, Paul. "Life at the Edge." *National Geographic* (June 2007). This article explores the shrinking sea ice at the North Pole and the effects it is having on the ecosystem.

Nii, Jenifer K. "Outdoor Industry Touts Its Clout." *Deseret News* (8/11/06). This article provides information on the significant contributions the outdoor recreation industry makes to the national economy.

O'Driscoll, Patrick. "Study Links Extended Wildfire Seasons to Global Warming." *USA Today* (7/7/06). Available online. URL: http://content.usatoday.com/community/tags/topic.aspx?req=tag&pg=202&tag=Earth. AccessedJune 29, 2009. Explores the cause-and-effect relationships of wildfire and recent findings and conclusions.

O'Hanlon, Larry. "U.S. Groundwater Drying Up." Discovery News (10/25/07). Available online. URL: http://dsc.discovery.com/news/2007/10/25/ground-water-drought-print.html. Accessed June 29, 2009. This article discusses the impacts global warming is having on the United States and the supplies of water that are necessary to support communities.

———. "Desalinated Water: Great to Drink, Bad for Crops." Discovery News (11/8/08). Available online. URL: http://dsc.discovery.com/news/2007/11/08/desalination-agriculture-print.html. Accessed June 29, 2009. This article discusses the issues surrounding desalination of ocean water in view of combating water shortages.

Owen, James. "Early Birds—Is Warming Changing U.K. Breeding Season?" National Geographic News (6/3/00). Available online. URL: http://news.nationalgeographic.com/news/2003/06/0603_030603_cheistmasowls.html. Accessed January 9, 2009. Explores the ecological effects of early spring warming.

Pickrell, John. "English Gardens Endangered by Warming." National Geographic News (2/10/03). Available online. URL: http://news.

nationalgeographic.com/news/2003/02/0210_030210_englishgarden.html. Accessed January 8, 2009. Discusses the changes in one of England's most popular pastimes.

———. "New Marine Conservation Area to Span Four Nations." National Geographic News (2/26/04). Available online. URL: http://google.nationalgeographic.com/search?site=default_collection&client=default_fr ontend&proxystylesheet=default_frontend&output=xml_no_dtd&oe= UTF-8&q=New+Marine+Conservation+Area+to+Span+Four+Nations&btnG .x=35&btnG.y=6. Accessed January 5, 2009. This looks at the creation of one of the world's largest marine protected areas.

Revkin, Andrew C. "Climate Experts Warn of More Coastal Building." *New York Times* (7/25/06). Available online. URL: www.nytimes.com/2006/07/25/science/earth/25coast.html. Accessed January 10, 2009. This article explores the economic and social issues of building in coastal areas that could be prone to hurricane damage.

———. "Global Warming Trend Continues in 2006, Climate Agencies Say." *New York Times* (12/15/06). Available online. URL: http://www.nytimes.com/2006/12/15/science/15climate.html. Accessed June 29, 2009. This article reports on new record-setting heat waves worldwide.

———. "A New Middle Stance Emerges in Debate over Climate." *New York Times* (1/1/07). Available online. URL: http://www.sourcewatch.org/index.php?title=Global_warming. Accessed June 29, 2009. This article introduces a new mindset on global warming; there exists not just a group of those who completely support the theory of global warming and a group who does not, but now a middle-of-the-road group that represents a more moderate mixture of the two extremes.

———. "Connecting the Global Warming Dots." *New York Times* (1/14/07). Available online. URL: http://www.nytimes.com/2007/01/14/weekinreview/14basics.html. Accessed June 29, 2009. This article presents the basics of what global warming is and why it is happening.

———. "U.S. Predicting Steady Increase for Emissions." *New York Times* (3/3/07). Available online. URL: http://www.nytimes.

com/2007/03/03/science/03climate.html. Accessed June 29, 2009. This article discusses information recently released from the White House predicting that emissions of carbon dioxide and other greenhouse gases are expected to increase over the next decade; citing emissions will grow 11 percent from 2002 to 2012 in the United States.

———. "Arctic Sea Ice Melting Faster, a Study Finds." *New York Times* (5/1/07). Available online. URL: www.nytimes.com/2007/05/01/us/01climate.html?ref=science. Accessed June 29, 2009. This article looks at the dilemma in the world's polar regions and discusses worldwide implications.

———. "Many Arctic Plants Have Adjusted to Big Climate Changes, Study Finds." *New York Times* (6/15/07). Available online. URL: http://www.nytimes.com/2007/06/15/science/15arctic.html. Accessed June 29, 2009. This article discusses the concept that some plants may be able to shift long distances to follow the climate conditions for which they are best adapted as those conditions move under the influence of global warming.

———. "Arctic Melt Unnerves the Experts." *New York Times* (10/2/07). Available online. URL: http://www.nytimes.com/2007/10/02/science/earth/02arct.html. Accessed June 29, 2009. This article presents information confirming that Arctic ice is melting faster than scientists originally expected and the global consequences of it.

———. "As China Goes, So Goes Global Warming." *New York Times* (12/16/07). Available online. URL: http://www.nytimes.com/2007/12/16/weekinreview/16revkin.html?fta=y. Accessed June 29, 2009. This article addresses the issue of clean energy for China.

———. "Issuing a Bold Challenge to the U.S. Over Climate." *New York Times* (1/22/08). Available online. URL: http://www.nytimes.com/2008/01/22/science/earth/22conv.html. Accessed June 29, 2009. This article explores the international political arena and the pressure being put on the United States to become more involved with addressing global warming issues.

Roach, John. "Penguin Decline in Antarctica Linked with Climate Change." National Geographic News (5/9/01). Available online. URL: http://news.nationalgeographic.com/news/2001/05/0509_penguindecline.html. Accessed June 29, 2009. Presents recent information about the plight of penguins suffering from the effects of a warmer environment.

———. "New Zealand Tries to Cap Gaseous Sheep Burps." National Geographic News (5/13/02). Available online. URL: http://news/nationalgeographic.com/news/2002/05/0509_020509_belch.html. Accessed January 9, 2009. Discusses the issues ranchers in New Zealand are having to face in light of the global warming problem.

———. "By 2050 Warming to Doom Million Species, Study Says." National Geographic News (7/12/04). Available online. URL: http://news.nationalgeographic.com/news/2004/01/0107_040107_extinction.html. Accessed June 29, 2009. Good overview on the potential effects of global warming to individual species in a range of different ecosystems.

Rosenthal, Elisabeth, and Andrew C. Revkin. "Panel Issues Bleak Report on Climate Change." *New York Times* (2/2/07). Available online. URL: http://www.nytimes.com/2007/02/02/science/earth/02cnd-climate.html. Accessed June 29, 2009. This article is a review of the IPCC's newly issued report on global warming.

Rosenthal, Elisabeth. "Science Panel Calls Global Warming 'Unequivocal.'" *New York Times* (2/3/07). Available online. URL: http://www.nytimes.com/2007/02/03/science/earth/03climate.html. Accessed June 29, 2009. This article profiles the IPCC's latest report on global warming.

———. "U.N. Report Describes Risks of Inaction on Climate Change." *New York Times* (11/17/07). Available online. URL: http://www.nytimes.com/2007/11/17/science/earth/17climate.html. Accessed June 29, 2009. This article focuses on the attention the United Nations is trying to gain in order to get individual countries to contribute toward the solution of global warming.

————. "U.N. Chief Seeks More Climate Change Leadership." *New York Times* (11/18/07). Available online. URL: http://www.nytimes.com/2007/11/18/science/earth/18climatenew.html. Accessed June 29, 2009. This article reviews the necessity of effective political leadership in order to back up the effort to stop global warming if it is to succeed.

Running, Steven W. "Is Global Warming Causing More, Larger Wildfires?" *Science* (8/18/06). Available online. URL: http://www.sciencemag.org/cgi/content/short/1130370. Accessed June 29, 2009. This article discusses the recent correlation between rising temperatures and the occurrence of wildfires.

Sabine, C. L., R. A. Feely, et al. "The Oceanic Sink for Anthropogenic CO_2." *Science* 305: 367–371 (1994). This article discusses the carbon sequestration role of the world's oceans as one storage reservoir of excess carbon.

Schirber, Michael. "Nature's Wrath: Global Deaths and Costs Swell." LiveScience (11/1/04). Available online. URL: http://www.livescience.com/environment/041101_disaster_report.html. Accessed June 29, 2009. This article examines the reasons why weather-related disasters are more prominent now than they were decades ago.

————. "Global Disaster Hotspots: Who Gets Pummeled." LiveScience (12/7/04). Available online. URL: http://www.livescience.com/environment/050107_disaster_hotspots.html. Accessed June 29, 2009. This article discusses the most likely areas in the world to be negatively affected by natural disasters.

————. "Longer Airline Flights Proposed to Combat Global Warming." LiveScience (1/26/05). Available online. URL: http://www.livescience.com/environment/050126_contrail_climate.html. Accessed June 29, 2009. This article investigates plans to limit the effects of contrails in the atmosphere from airline traffic.

Schmid, Randolph E. "Greenhouse Gas Hits Record High." LiveScience (3/15/06). Available online. URL: http://www.planitbuildit.com/scientificproof.htm. Accessed June 29, 2009. This article discusses how the measurements of atmospheric carbon dioxide have changed, especially since the beginning of the Industrial Revolution.

———. "Global Warming Differences Resolved." *LiveScience* (5/2/06). Available online. URL: http://www.usatoday.com.tech.science/2006-05-02-warming-temps-resolved_x.htm. Accessed June 29, 2009. This article discusses how former discrepancies in temperature calculations between satellite and radiosonde data used for global warming analysis have finally been identified and corrected.

ScienceDaily. "From Icehouse to Hothouse: Melting Ice and Rising Carbon Dioxide Caused Climate Shift." (2/27/07). Available online. URL: http://www.sciencedaily.com/releases/2007/02/0702211358.htm. Accessed January 1, 2009. This article discusses modeling of paleoclimates in order to understand changes happening today.

Silverman, Fran. "What's a Ski Area to Do as It Warms? Adapt." *New York Times* (1/13/08). Available online. URL: http://www.nytimes.com/2008/01/13/nyregion/nyregionspecial2/13skict.html?fta=y. Accessed June 29, 2009. This article looks at the impacts on the recreation industry as a result of global warming.

Simons, Marlise. "Fungus Once Again Threatens French Cave Paintings." *New York Times* (12/9/07). Available online. URL: http://www.nytimes.com/2007/12/09/world/europe/09cave.html. Accessed June 29, 2009. This article looks at the cultural effects of global warming and why the world's priceless art may be in danger.

Space Daily. "Global Warming to Squeeze Western Mountains Dry by 2050." (2/18/04). Available online. URL: http://www.spacemart.com/reports/Global_Warming_To_Squeeze_Western_Mountains_Dry_By_2050.html. Accessed January 11, 2009. This article addresses the effects global warming will have on future snowpack and how that will impact the use and demand of water.

Stevens, William K. "On the Climate Change Beat, Doubt Gives Way to Certainty." *New York Times* (2/6/07). Available online. URL: http://www.nytimes.com/2007/02/06/science/earth/06clim.html. Accessed June 29, 2009. As research continues, there is enough evidence that scientists can no longer say the climate is not warming up.

Struck, Doug. "NOAA Scientists Say Arctic Ice Is Melting Faster Than Expected." *Washington Post* (9/7/07). Available online. URL: http://www.washingtonpost.com/wp-dyn/content/article/2007/09/06/AR2007090602499.html. Accessed June 29, 2009. This article offers evidence that predictions of future ice loss may be too conservative.

Sturm, Matthew, Donald K. Perovich, and Mark C. Serreze. "Meltdown in the North." *Scientific American* (October 2003). This article looks at the current melting in the Arctic regions, what it is doing to the ecosystems, and how that will affect the rest of the world.

Than, Ker. "Animals and Plants Adapting to Climate Change." LiveScience (6/21/05). Available online. URL: http://www.livescience.com/environment/050621_warming_list.html. Accessed June 29, 2009. This looks at how some species in ecosystems have been able to adapt to changing systems.

———. "How Global Warming Is Changing the Wild Kingdom." LiveScience (6/21/05). Available online. URL: http://www.livescience.com/environment/050621_warming_changes.html. Accessed June 29, 2009. This article looks at ecological issues among mammals, fish, insects, and other types of wildlife.

———. "The 100-Year Forecast: Stronger Storms Ahead." LiveScience (10/13/05). Available online. URL: http://www.livescience.com/environment/051013_stronger_storms.html. Accessed June 29, 2009. This article presents the concept of an "enhanced" water cycle and more severe storms as a result of global warming in the future.

———. "Polar Meltdown Near: Seas Could Rise 3 Feet Per Century." LiveScience (3/23/06). Available online. URL: http://www.livescience.com/environment/060323_ice_melt.html. Accessed June 29, 2009. This article explores the implication of sea level rise as a result of the melting of polar ice caps.

———. "Global Warming Weakens Pacific Trade Winds." LiveScience (5/3/06). Available online. URL: http://www.msnbc.msn.com/id/12612965/. Accessed June 29, 2009. This article outlines why the trade winds are weakening as a result of global warming.

Tibbetts, Graham. "Global Warming Could Close Half of Alpine Ski Resorts by 2050." Climate Ark (12/3/03). Available online. URL: http://www.climateark.org/shared/reader/welcome.aspx?linkid=27459. Accessed June 29, 2009. This presents the current situation facing ski resorts in Europe.

Tierney, John. "In 2008, a 100 Percent Chance of Alarm." *New York Times* (1/1/08). Available online. URL: http://www.nytimes.com/2008/01/01/science/01tier.html. Accessed June 29, 2009. This article presents an overview of why action needs to be taken immediately to deal with global warming on a global scale.

Time. "Global Warming: The Causes, The Perils, the Politics—and What It Means for You." (4/9/07). Available online. URL: http://www.alibris.com/search/books/qwork/10358045/used/Time:%20Global%20Warming:%20The%20Causes%20-%20The%20Perils%20-%20The%20Solutions%20-%20The%20Actio ns:%2051%20Things%20You%20Can%20Do. Accessed June 29, 2009. This article suggests 51 ways to save the environment and curb global warming.

Trivedi, Bijal P. "Mosquito Adapting to Global Warming, Study Finds." National Geographic News (11/5/01). Available online. URL: http://news.nationalgeographic.com/news/2001/11/1105_Tvmozzie.html. Accessed June 29, 2009. This article describes how mosquitoes are currently migrating and adapting to warmer temperatures.

———. "South African Desert Becomes Global Warming Lab." National Geographic News (8/4/03). Available online. URL: http://news.nationalgeographic.com/news/2003/08/0804_030804_karoo.html. Accessed June 29, 2009. This article looks specifically at one desert and analyzes the effects global warming is having now and outlines what is predicted for its future.

USA Today. "Global Warming Threatens Alpine Ski Resorts." (12/13/06). Available online. URL: http://www.swissinfo.org/eng/front/detail/Climate_change_threatens_ski_resorts_in_Europe.html?siteSect=105&sid=7347238&cKey=1166083840000. Accessed June 29, 2009. Discusses current economic problems the ski resorts are facing in the European Alps due to global warming.

Vergano, Dan. "Global Warming Stoked '05 Hurricanes, Study Says." *USA Today* (6/25/06). Available online. URL: http://www.usatoday. com/tech/science/2006-06-22-hurricane-blame_x.htm. Accessed June 29, 2009. This article presents the reasons why climate scientists believe that global warming helped fuel 2005's destructive hurricane season.

Wald, Matthew L. "Cleaner Coal Is Attracting Some Doubts." *New York Times* (2/21/07). Available online. URL: http://www.nytimes. com/2007/02/21/business/21coal.html. Accessed June 29, 2009. This article explores the use of new technology called 'gasification' to operate new coal plants and whether it will be as environmentally friendly as originally thought.

———. "Study Details How U.S. Could Cut 28% of Greenhouse Gases" *New York Times* (11/30/07). Available online. URL: http://www.sehn. org/tccgreenhousegasses.html. Accessed June 29, 2009. This article outlines how the United States could reduce the greenhouse gases it generates at a reasonable cost with only small technological innovations.

Walsh, Bryan. "The Fire This Time." *Time* (10/25/07). Available online. URL: www.time.com/time/nation/article10,8599,1675380,00. html. Accessed February 23, 2009. This article touches on wildfires in California and its connections to climate change.

Warburton, Louise. "No Where Else to Go—Climate Changes and Parrots: Can they Adapt to Survive?" *Bird Talk* (January 2007). This article discusses the ecological impact global warming is having on the world's bird population.

Williams, Gisela. "Resorts Prepare for a Future Without Skis." *New York Times* (12/2/07). Available online. URL: http://travel.nytimes. com/2007/12/02/travel/02skiglobal.html. Accessed June 29, 2009. This article discusses the economic ramifications of global warming on the tourist industry.

Williams, Jack. "Drilling Uncovers Past, Maybe the Future." *USA Today* (1/23/99). This article reviews what climatologists have learned about climate change through the study of climate from the past via ice cores.

————. "Greenland's Ice Tells of Past Climates, Maybe Ancient Life." *USA Today* (8/30/04). This article reviews the importance of paleoclimatology.

Woods Hole Oceanographic Institution. "Abandoned Walrus Calves Reported in the Arctic." *Oceanus Magazine* (2/11/08). Looks at the dilemma walrus face due to the massive melting of Arctic sea ice and the predicament it puts their ultimate survival in.

Woollard, Rob. "California Wildfires Force Mass Evacuations." Discovery News (10/23/07). Available online. URL: http://dsc.discovery.com/news/2007/10/23/california-wildfire-print.html. Accessed June 29, 2009. This article links the recent tragic wildfires of California and the American Southwest with drought and global warming.

Zimmer, Carl. "Migration, Interrupted: Nature's Rhythms at Risk." *New York Times* (1/1/08). This article discusses migration corridors and how global warming is affecting them.

WEB SITES

Global Warming

Climate Ark home page. Sponsored by Ecological Internet. Available online. URL: www.climateark.org. Accessed June 29, 2009. A Web site that promotes public policy that addresses global climate change through reduction in carbon and other emissions, energy conservation, alternative energy sources, and ending deforestation.

Climate Solutions home page. Sponsored by Atmosphere Alliance and Energy Outreach Center. Available online. URL: www.climatesolutions.org. Accessed June 30, 2009. A Web site that offers practical solutions to global warming.

Environmental Defense Fund home page. Sponsored by Environmental Defense Fund. Available online. URL: www.environmentaldefense.org. Accessed June 30, 2009. A Web site of an organization started by a handful of environmental scientists in 1967 that provides quality information and helpful resources on understanding global warming and other crucial environmental issues.

Environmental Protection Agency home page. Sponsored by the U.S. Environmental Protection Agency. Available online. URL: www.epa.gov. Accessed June 30, 2009. This Web site provides information about EPA's efforts and programs to protect the environment. It offers a wide array of information on global warming.

European Environment Agency home page. Sponsored by the European Environment Agency in Copenhagen, Denmark. Available online. URL: www.eea.europa.eu/themes/climate. Accessed June 30, 2009. This Web site posts their reports on topics such as air quality, ozone depletion, and climate change.

Global Warming: Focus on the Future home page. Sponsored by EnviroLink. Available online. URL: www.enviroweb.org. Accessed June 30, 2009. This Web site offers statistics and photography of global warming topics.

HotEarth.Net home page. Sponsored by National Environmental Trust. Available online. URL: www.net.org/warming. Accessed June 30, 2009. This Web site features informational articles on the causes of global warming, its harmful effects, and solutions that could stop it.

Intergovernmental Panel on Climate Change (IPCC) home page. Sponsored by the World Meteorological Organization (WMO) and the United Nations Environment Programme (UNEP). Available online. URL: http://www.ipcc.ch/. Accessed June 30, 2009. This Web site offers current information on the science of global warming and recommendations on practical solutions and policy management.

NASA's Goddard Institute for Space Studies home page. Sponsored by the National Aeronautics and Space Administration. Available online. URL: www.giss.nasa.gov. Accessed June 30, 2009. This Web site provides a large database of information, research, and other resources.

NOAA's National Climatic Data Center home page. Sponsored by the National Oceanic and Atmospheric Administration. Available online. URL: www.ncdc.noaa.gov. Accessed June 30, 2009. This Web site offers a multitude of resources and information on climate, climate change, global warming.

Ozone Action home page. Sponsored by the Southeast Michigan Council of Governments. Available online. URL: www.semcog.org/OzoneAction.aspx. Accessed June 30, 2009. This web site provides information on air quality by focusing on ozone, the atmosphere, environmental issues, and related health issues.

Scientific American home page. Sponsored by Scientific American, Inc. Available online. URL: www.sciam.com. Accessed June 30, 2009. This organization offers an online magazine and often presents articles concerning climate change and global warming.

Tyndall Centre at University of East Anglia home page. Sponsored by the Tyndall Centre for Climate Change Research. Available online. URL: http://www.tyndall.ac.uk. Accessed June 30, 2009. This Web site offers information on climate change and is considered one of the leaders in UK research on global warming.

Union of Concerned Scientists home page. Sponsored by the Union of Concerned Scientists. Available online. URL: www.ucsusa.org. Accessed June 30, 2009. This Web site offers quality resource sections on global warming and ozone depletion.

United Nations Framework Convention on Climate Change (UNFCCC) home page. Sponsored by the United Nations Framework Convention on Climate Change. Available online. URL: http://unfccc.int/2860.php. Accessed June 30, 2009. This Web site resents a spectrum on climate change information and policy.

U.S. Global Change Research Program home page. Sponsored by the U.S. Office of Science and Technology Policy, the Office of Management and Budget, and the Council on Environmental Quality. Available online. URL: www.usgcrp.gov. Accessed June 30, 2009. This Web site provides information on the current research activities of national and international science programs that focus on global monitoring of climate and ecosystem issues.

World Wildlife Foundation Climate Change Campaign home page. Sponsored by the World Wildlife Fund. Available online. URL: www.worldwildlife.org/climate/. Accessed June 30, 2009. This Web site contains information on what various countries are doing, and not doing, to deal with global warming.

Greenhouse Gas Emissions

Energy Information Administration home page. Sponsored by the U.S. Department of Energy. Available online. URL: www.eia.doe.gov/environment.html. Accessed June 30, 2009. This Web site lists official environmental energy-related emissions data and environmental analyses from the U.S. government. This site contains U.S. carbon dioxide, methane, and nitrous oxide emissions data and other greenhouse reports.

World Resources Institute—Climate, Energy & Transport homepage. Sponsored by the World Resources Institute. Available online. URL: www.wri.org/climate/publications.cfm. Accessed June 30, 2009. This Web site offers a collection of reports on global technology deployment to stabilize emissions, agriculture, and greenhouse gas mitigation, climate science discoveries, and renewable energy.

INDEX

Italic page numbers indicate illustrations or maps. Page numbers followed by *c* denote entries in the chronology.

A

Abisko, Sweden 100
acidification. *See* ocean acidification
adaptation
    forest ecosystems 66, 75–76
    future issues 199–200
    grassland ecosystems 31, 90
    marine ecosystems 196–197
    polar ecosystems 116–119
    tundra loss 111
Africa 128
Agapornia nigrigenis 72–73
Agrawala, Shardur 59, 161
agriculture 67, 70, 89–90, 124
Alaska 13, 92, 100, 150
albedo 11
algae 12, 187–188, 196
algae blooms 176–177
Allen, Myles 40
alpine zones 158–159
Amazon rain forest 68–69, 140
Antarctic ecosystems 111–116, *114*
Antarctic Ice Sheet 14–15, 205*c*, 207*c*
Antarctic Peninsula 111
Antarctic sea ice 115
anthropogenic climate change 3, 8, 101, 181
Appalachian Range 56
aquaculture 175–176
aquatic environments. *See* marine ecosystems
Arctic ecosystems 92–111
    global warming impact 95–109
    permafrost 109–110
    shifting vegetation zones 110–111
    wildlife 100–109, *102, 107*
Arctic ice cap 93, 111
Arctic ice pack 106
Arctic lakes/ponds 95
Arctic sea ice *97*
    in Arctic ecosystems 92–95
    as early warning system for global warming 97–98

polar bears and 103–104
    retreat of 8–9, 92–94
    seals and 107–108
artificial snow 59–60
Ashjian, Carin 106
Asian rangelands 86–87
Asian steppes 31
Asner, Greg 70
Atelopus varius 18
Australia 14, 73, 112–114, 125, 153–154, 188
Austria 162
Azeez Abdul Hakeem, Abdul 189–190

B

Bakum, Andrew 175
Bali Road Map 183
Beardall, John 184
Beckett, Margaret 40
Belize 190
Bengal tiger 25, *87*
Beniston, Martin 154–155
Bhutan 157
biodiversity 21–35, *24–26, 28,* 56–57, 133–135, 144
biofuels, seaweed as 184
birds 30, 60–61, 95, 100, 172, 193
black-cheeked lovebird 72–73
Blair, Tony 40
bleaching (coral) 3, 12, 186, 190
blue-green algae 196
boreal forests 30, 50, 55, 65–67, *66,* 98, 100
Botswana 128
Boyes, Steve 74
Bradshaw, William 62
Brigham-Grette, Julie 111
Brinson, Mark 196
Bromus tectorum. See cheatgrass
brook trout 166
budgies 73, *74*
buffer zones 41
Bufo periglenes 18
Burki, Rolf 162
burrowing parrot 73–74
Bush, George W. 190, 208*c*